SHORT STUDIES
ON GREAT SUBJECTS

By J. A. Froude

*Selected and with an Introduction
by David Ogg*

Short Studies on Great Subjects

By J. A. FROUDE

Selected and with an Introduction by
DAVID OGG

Cornell Paperbacks

CORNELL UNIVERSITY PRESS

ITHACA, NEW YORK

PRINTED IN THE UNITED STATES OF AMERICA

BY VALLEY OFFSET, INC.

CONTENTS

INTRODUCTION

Talleyrand is said to have remarked that only those who had experience of existence before 1789 really knew how pleasant life could be. For us the limiting date is 1914. Thus the Victorians avoided the Deluge by a comfortable margin. As we look back on their ideals and achievements we may feel that the gap between their age and ours is not to be measured by years, still less by advances in science and technology, but by something greater and more imponderable —by the change from conviction to doubt and disillusionment. Our Victorian ancestors, protected by the Channel from the repercussions of continental politics, and spared from menial tasks by a scarcely known population 'below stairs', could devote themselves to serious study and meditation without the distractions of telephone, radio and television, secure in an almost insulated world. The sense of duty was the most acclaimed of all the senses; life had both purpose and direction, moral standards were uncompromising. Our attitude to that age has often been one of patronage or condescension, but it may well be mingled with a feeling of envy.

Such a milieu is clearly reflected in the career and character of James Anthony Froude[1] (1818-1894). He was the second son of Robert Hurrell Froude, rector of Dartington in Devon and afterwards archdeacon of Totnes, a strict churchman who imposed somewhat austere ideals on his family. After three years as a scholar at Westminster School—where, as in all the public schools of that time life was hard and harsh—the young Froude matriculated at Oriel College, Oxford in December 1835 where he had rooms on the same staircase as Newman. In this way the Devonshire youth was, at an impressionable age, brought into the midst of men afterwards

[1] There are two biographies—Herbert Paul, *The Life of Froude*, London 1905, and Waldo Hilary Dunn, *J. A. Froude, a Biography*. Of the latter, the first volume, covering the period 1818-1856 was published by the Clarendon Press in 1961.

to be known as the Tractarians; but, unlike his elder brother Hurrell Froude, James Anthony soon reacted against the movement, and this revival of Catholicism within the Church of England proved, in later life, to be one of the things that aroused his passionate resentment. In 1840 he obtained honours in the school of Literae Humaniores, a course of study based on ancient history and philosophy, studied in the originals, a discipline still held in highest esteem by the University of Oxford because it provides the soundest basis for the student who intends to devote himself to other humane subjects, such as history or literature. Both in its immense range and its evidence of keen critical faculty Froude's literary work serves to confirm this assumption. It was natural that he should think of an academic career. At that time most of the college fellowships were tenable only by· those who had proceeded to at least the first stage in holy orders; accordingly, on his election to a fellowship at Exeter College, Froude, in 1844, took deacon's orders. Five years later he published his *Nemesis of Faith* which was regarded by some readers as a criticism of the psychological effects of the Oxford Movement. A copy of the book was publicly burned in the hall of Exeter College by the Rector, Dr. Sewell, and immediately Froude resigned his fellowship. In 1872 a change in the law enabled him to divest himself of holy orders. Not until 1892, when he was appointed Regius Professor of Modern History, did Froude return to Oxford in an official capacity.

While supporting himself by literary work and teaching, Froude made a number of important friendships, including that with Charles Kingsley (whose sister-in-law he married), Arthur Hugh Clough, Matthew Arnold and James Spedding. It was Spedding who introduced him to Carlyle in 1849, an event followed by the most formative but not perhaps the most fortunate friendships of Froude's life. From Carlyle he derived the ' great man ' interpretation of history, according to which evolution is determined, not by social or economic forces, but by the domination of dynamic personalities— including Frederick the Great and Henry VIII. Admiration for German philosophy and institutions was one more of the debatable influences impressed on his disciple by the sage of Chelsea. In later years this friendship was to have even more unfortunate consequences. In 1881, when Carlyle died, Froude was the sole surviving literary executor, a capacity which gave him access to the most intimate details of Carlyle's married life, a life which had been none too happy for Jane Welsh Carlyle. It appears that after her death (in 1866) the widower desired to make some kind of public penance for his harshness or presumed harshness to her in their married life, but on this point the instructions to the executors were neither clear nor consistent. Between 1881 and 1884 Froude devoted a number of volumes to this unpromising subject, including two biographies and a volume of letters and memorials of Jane Welsh Carlyle. The reception accorded to these publications supports the view that one should not write a biography of one's hero until all his friends and relations are dead. In his frankness—sometimes brutal frankness—Froude gave great offence to many who had known the two Carlyles; still more, he was justifiably accused of serious mistakes in the transcription of his literary originals. Froude always had a burning zeal for truth; but, in our close-knit civilisation, this zeal must be tempered by caution or restraint, and the evidence must be incontrovertible. By this time Froude had won for himself many critics; the Carlyle volumes greatly added to their number.

But this is to anticipate. Froude supported himself by the publication of books and by contributions to literary journals, including Fraser's Magazine, of which he was editor from 1860 to 1874. Some of his essays were reprinted in *Short Studies on Great Subjects,* of which the first series was published by Longmans in 1867 and the fourth in 1883. But in these middle years of his life he was occupied mainly with his *History of England*, covering the period from the death of Wolsey to the defeat of the Spanish Armada, of which the first two volumes were published in 1856 and the last in 1870. Several editions of the twelve volumes were published in his lifetime. These dates nearly coincide with those of the earlier editions of Macaulay's *History,* and thus the two most widely read contributions to English History were produced in the high noon of the Victorian day. Though dealing with different periods and often contrasted in literary style, both books have in common a detestation of what is called Popery and a fervent belief in Protestantism—not as a set of theological opinions, but as a way of life, in pursuit of which England's greatness had been achieved. Another important element shared by these two historians is a strong sense of drama in the stirring events of national achievement. Both of them were great writers of narrative because they could visualise what they were describing, and could infuse into their pages something of the emotional intensity by which they were inspired.

These qualities are outmoded in the serious historiography of to-day. We have lost the confidence and the convictions of the Victorians, and on the crowded highways of history writing, Safety First is now the motto. For the problems of the past as for those of the present we put our faith in committees—hence the abundance of composite volumes of history in which the separate contributions are carefully checked and pruned so as to produce a uniform pattern. In trying to eliminate the personal element from the writing of history we often achieve no more than the non-committal. It is not surprising therefore that Macaulay and Froude have fared badly in more than a century of criticism, much of it niggling, some of it vindictive. But on one count both can be acquitted. Neither was drawing on his imagination for the facts; both spent years of hard work in the study of original material, much of it in manuscript. Froude made exhaustive use of the archives at Simancas, devoting many months to transcription of documents, some of them almost illegible. His transcripts were deposited in the British Museum, so that his use of his sources can be checked. No less an authority than the late Professor Pollard has vouched for the substantial accuracy of Froude's narrative. Moreover, it is often forgotten that if one writes history on the Froude-Macaulay scale, the number of possible errors—mostly venial—can be numbered not in tens, but in hundreds. Is it surprising that we no longer write large-scale, one-man histories? The risk, like the labour, is too great.

Froude led a busy, active life, in the course of which he travelled widely. Ireland he knew intimately from frequent visits and prolonged stays in that country. His views on Irish problems were considered controversial, but who has written seriously on Ireland without arousing controversy? Another set of problems was presented by South Africa, which he visited in 1874-5 in a semi-official capacity in order to explore the possibility of confederation. He was vehement in his condemnation of the British confiscation of the Griqualand diamond fields (1871) and he spoke, not always with discretion, on the desirability of federation. But he was regarded as no more than the mouthpiece of Lord Carnarvon's colonial policy, and was faced with the objection that proposals for changing the relations of the South African republics must originate not from London but from the states themselves. In general, and in opposition to the Little Englanders, Froude believed that the colonies had a great future and that emigration from our over-populated island should be encouraged. In 1884-5 he was in Australia, where he does not appear to have originated any controversy, but in 1886-7, when in the West Indies, he aroused criticism by his public scepticism about the fitness of the Caribbean possessions for representative government, and his views on the negroes were considered reactionary. Scotland he knew from frequent visits to Edinburgh where he served as a commissioner for the Scottish Universities in 1876.

In 1892 the death of E. A. Freeman, Regius Professor of History at Oxford created a vacancy which Froude was invited by Lord Salisbury to fill. It was a measure of justice that Froude should now occupy the chair formerly occupied by one of his most unsparing and even unfair critics; it was also proof of real achievement that he should return in such a distinguished capacity to a university which, more than forty years before, he had left in an atmosphere almost of opprobrium. As professor he lectured on Erasmus, on the Council of Trent and on English Seamen in the Sixteenth Century. The last of these, published in a single volume, is one of the best of his books. He was proud of his native Devon, and of the great seamen whom it has produced;

always he preferred men of action to thinkers and scholars. He loved the sea and sailed his own yacht. Few historians have done so much justice to the maritime elements in English history as Froude.

He died in his native county on Oct. 20 1894 and was buried at Salcombe. In spite of the many controversies in which he had been engaged and in spite of a certain aloofness and sensitiveness of temperament he impressed those who knew him by the genuineness of his character, the great range of his knowledge and experience and the brilliance of his conversation. The portraits[2] of him suggest a certain austerity and even asceticism, mingled with a vein of the sardonic, such as helps to account for his cult of Lucian, the satirist of the second century A.D., whose brilliant exposure of credulity and quackery provided the theme of ' A Cagliostro of the Second Century ', in one of the Short Studies. A reading of the Short Studies suggests that, of all his contemporaries, none showed such variety and range of literary output—extending from the remote problems of biblical history, the episodes and civilisation of ancient Greece and Rome, the ideals and struggles of medieval Europe, the achievements of the Protestant Reformation to the questions of his own day. In this long range the only obvious gaps are the seventeenth and eighteenth centuries. Their omission may have been entirely accidental. On the other hand Froude had no great interest in economic or constitutional history; his canvas was a large one, in which he sought to depict the clash of great personalities. He may have been alienated by the bitter religious strife of the seventeenth century which appeared to dash the hopes raised by the Elizabethan age. In the eighteenth century he may have found a formalism and even insincerity which alienated him; to the subterranean avenues of party politics he was indifferent. His devotion was to great causes and to the achievements of great men, particularly men of action. In this way his writings may provide an antidote to surfeit and disillusionment.

The selected essays here re-printed have been arranged

[2] Reproductions will be found in the two biographies referred to in the footnote, p. 7.

according to the chronological order of their subject matter, except that the first, the Science of History, does not relate to any one period. Since the days of Vico and Montesquieu attempts have been made to find principles determining the course of history, attempts which have usually been unfortunate, and for reasons expounded by Froude himself. In this, the first of the essays here reproduced, he took for his text the *History of Civilisation in England* by Henry Thomas Buckle (1821-1862). A believer in the determining influence of material conditions, including climate, Buckle thought that the course of human evolution could be reduced to principles as ' valid ' as those of Political Economy; he suggested also that the disturbing factor of human eccentricity could be eliminated by some kind of law of averages. It is no derogation to Buckle to say that he was largely a self-taught man; on the other hand, if he had had more contacts with the learned world, he might have developed a keener critical faculty, and might have learned that the history of thought is strewn with rejected hypotheses.

Froude dealt gently with Buckle. He showed that History can never be a science in the sense that the term is applied to, say, Astronomy, because in the latter there are regularly recurring phenomena which make possible both deduction and accurate prediction. As for an average of human conduct this varies from generation to generation. Freedom of the human will must always create an imponderable and unpredictable element. Few would disagree with these criticisms. On a broad view, according to Froude, it can be maintained that the world is built on moral foundations—the view that evil finally defeats itself and that ultimately, after much suffering, good prevails. But this is not a science. Froude was probably the most lenient of those who have criticised Buckle.

From these broad generalisations a welcome relief is provided by the second essay : Society in Italy in the Last Days of the Roman Republic. Here we have a minute and intricate study of a small fraction of the actual doings of humanity— a fast moving drama of real life as enacted in the Neapolitan town of Larino in the first century B.C. What shocks the reader is not so much the crimes of Caius Oppianicus, the

villain of the piece, as the corruption of the Roman administration of justice. Froude derived his facts mainly from
Cicero's two speeches *Pro Cluentio*. In so far as this essay has
a purpose it is to discredit the popular theory that, even in its
last days, the Republic was more moral than the Empire which
succeeded. By unravelling the tangled skein of Cicero's two
speeches Froude showed both patience and ingenuity, his essay
has its value as a ' close-up ' of private life in ancient times.
Its subject is sharply contrasted with the third study here
reproduced—a Bishop of the Twelfth Century, an account of
the career of St Hugh of Avalon who was bishop of Lincoln
from 1186 to 1200. Though not a professed medievalist
Froude was strongly attracted to the spiritual element in the
life of the middle ages, and in St Hugh he found an
ecclesiastic after his own heart. In contrast with Becket, to
whom Froude devoted one of the Short Studies, St Hugh
had a sense of humour; he was as resolute as his martyred
predecessor in defence of the Church's rights, but he had
humanity, as shown in his treatment of the poor. An even
more rare quality—he did not believe all the stories of miracles
that were brought to his notice. Happy in his subject, Froude
was here at his best.

His interest in medieval saints—their asceticism struck a
sympathetic chord in his nature—is reflected also in the
fourth essay : The Lives of the Saints, but here the attitude
is more critical. In the seventeenth century the Belgian
Jesuit J. Bolland had devoted himself to the compilation of a
complete calendar of the saints of western Christendom,
arranged according to the days of the month, but he died
before completing the month of March. Numerous con-
tinuators carried on the work. The Oxford Movement revived
the cult of hagiography in England and, in his early days,
Froude himself had been induced by Newman to contribute
a life of St Neot to *The Lives of the English Saints* (1844-5);
an enterprise which may well have aroused scepticism in a
mind so critical as Froude's. He concluded that, on a slender
basis of fact, indeed often on no facts at all, vast legends might
be built up, each generation contributing its quota of miracles.
Hence, if we must have hagiography, we should prefer the

earlier to the later lives, as the former, though less literary, are less fanciful and often have a germ of poetic truth. As for ' facts ', ' we cannot ', writes Froude, ' relate facts as they are; they must first pass through ourselves, and we are more or less than mortal if they gather nothing in the transit '. This can be applied to all historical writing. But Froude was not a mere iconoclast. However unauthenticated the recorded lives of the saints may be, they have their value as embodying ideals of the Christian life—in them ' the Catholic mind was expressing its conception of the highest human excellence '. They provided for the medieval mind a form of hero worship, and it is a poor age that is completely lacking in such ideals.

In the fifth study, the Dissolution of the Monasteries, we have reached the sixteenth century, a period which Froude had made his own. But he was now well within the zone of controversy. The monasteries were dissolved on the allegation (fully substantiated in many cases) that they had fallen into hopeless corruption and were no longer serving the purposes for which they had been founded. Here is the main issue on which battle has been joined by Catholic and Protestant. In recent years the most formidable contestants were Cardinal Gasquet and Dr Coulton; nor has the dust raised by their polemics had time to settle. We are all aware how foolish it is to argue from the particular to the general; could we select our monasteries, we could make a case either way. Apart from this, however, it is possible that the contestants have read modern ideas into the past. To-day, whatever our differences, we insist on linking religion with morality, but this association did not always exist in the Middle Ages because, in spite of high ideals, people thought of religious duty in terms more concrete, or even more materialist. Increasing exemptions and dispensations, together with proficiency in a subtle dialectic, made it easy to evade and even to pervert the strict injunctions of medieval founders, and it is almost a commonplace to say that religious institutions degenerated in direct proportion to the exalted aims of their original intention. This process was intensified in late medieval times, a fact recognised as much by Catholics as by Protestants, and it led directly to the reforms of the Council of Trent. The controversy over the Dissolution has been so bitter because expressed in terms of black and white.

Study number six is in happier vein. England's Forgotten Worthies treats of the men who, more than any others, stirred our author's admiration—the English seamen of the sixteenth century. This enthusiasm may account for the high literary level of the essay. These rough sailors were inspired by something more than the lust for gain; they were moved by the iniquities committed by the Spanish Inquisition on their compatriots. This was not an abstract hatred of Roman Catholicism, but a fierce detestation of injustice and brutality. Moreover it was in this mould of hatred that English nationalism took shape. As Protestant, Patriot and Imperialist Froude, for perhaps the first time, showed the part played by the sixteenth-century seamen in the evolution of the English nation.

A mausoleum is not usually regarded as a suitable source of inspiration, but in the seventh study here presented: Cheneys and the House of Russell, Froude succeeded in focusing our attention on the Russell family from the standpoint of their burial place at Cheneys in Buckinghamshire. The chapel was built by Anne, Countess of Bedford in 1556, a year when it seemed possible that Protestantism would be extirpated from the land. It was to Henry VIII that the Russell family owed the beginning of its fortunes, and there can be few families which, in widely scattered periods, have played such a distinguished and usually honourable part in our public life. The founder of the family, Sir John Russell, had been active in the suppression of the monasteries, and his descendants managed to retain office even in the troubled times which followed the death of Henry VIII. Of the more notable members of this lavishly endowed House one may note Lord William Russell who, on an accusation of complicity in the Rye House Plot, was executed in 1683, so providing the Whig cause with a martyr; and it was as Whigs that the family exercised its influence in the affairs of the nation. The naval battle which destroyed the hopes of the exiled James II and ensured the success of the Protestant Revolution was won in 1692 by Admiral Russell at La Hogue. As Dukes of Bedford the Russells were among the most powerful of those families which dominated politics in the eighteenth century, and it was of one of them that Burke said : ' I was not, like His Grace of Bedford, swaddled and rocked and dandled into a legislator '. A more distinguished role was played by Lord John Russell (1792-1878), a leader of the movement for parliamentary reform who, in 1828, obtained the repeal of the Test and Corporation Acts, and in December 1831 (after two abortive attempts) successfully moved the adoption of the Reform Bill which became law in the following year. So Froude's study of a mausoleum led naturally to an appreciation of the part played in our history by successive generations of a great family.

The eighth study returns to a topic somewhat more controversial—what Froude calls : The Oxford Counter Reformation, of which letters III and IV are here re-printed. Of these, letter III is an appreciation of John Henry Newman, and letter IV, somewhat more critical, deals with the famous Tract XC and its consequences. If a man is to be known by his friendships, then it may be sufficient to say that Froude's two greatest friends were Newman and Carlyle—the most ill-assorted pair in the century. ' The hearts of men vibrate in answer to one another like the strings of musical instruments ', such was our author's explanation of the extraordinary personal influence exercised, particularly in his sermons, by the founder of the Oxford Movement; ' a sermon from him was a poem, formed on a distinct idea, fascinating by its subtlety '. The intensity and honesty of Newman's faith perplexed Froude who could only attribute it to some extra sense not accorded to all men. Moreover, the fascination exercised by Newman was scarcely dimmed with time, and Froude's appreciation is probably the most convincing of all that were voiced by his contemporaries. Letter IV is concerned with Tract XC in which Newman attempted to reconcile two things which are probably quite incompatible—the Catholicism of the medieval English Church and the Protestantism of the Articles of Religion formulated in the Elizabethan settlement. As the Anglican clergy had sworn acceptance of these Articles, it was clear that those of them who were drawn into the Oxford Movement were in a position of acute difficulty. Froude had some sympathy with Newman's attempted reconciliation, but it has to be confessed that the attempt savours of sophistry. But, whatever may be true of the sixteenth century, it was the Revolution of 1688 which made impossible the attempt to bridge the gap. The rule of the later Stuarts proved that there could be no *via media*.

The last two studies in this selection—On the Uses of a Landed Gentry and On Progress provide good examples of Froude's forthright views on the life around him. He believed that a landed gentry might be a powerful influence for good, and cited the example set by Augustus John Smith (1804-1872) who, in 1834, obtained from the crown a lease of the Scilly Isles. A benevolent despot, Smith devoted himself to the best interests of the islanders, suppressing improvidence and idleness by methods considered high-handed, but succeeding in the creation of a spirit of independence and thrift. Froude expressed a preference for large estates, as contrasted with small ones, since, in the former, there is usually enough capital to provide the tenant with adequate facilities. In this respect he contrasted conditions in England with those prevailing in Ireland—where there was the curse of absenteeism—and on the continent where, as so many landowners lived in towns, there was so little of that personal contact with the tenantry so characteristic of the best type of English landowning. Froude maintained that the great landlords were restrained in the exercise of their powers by the force of public opinion; they have since been restrained by something even stronger, namely, taxation. This study concludes with an appeal for more emigration to the colonies as an outlet for excess population at home.

The last essay—that on Progress—is perhaps the most controversial of all included in this volume. It provides a proof that not all the Victorians were proud or even complacent about the changes going around them, and some readers may find in this essay evidence of a reactionary temper. Thus, he had no belief in the popular vote, maintaining that parliament in the eighteenth century was a better nursery of statesmen. He was sceptical about admission to the Civil

Service by open competition; indeed he appears to have had his doubts about the state system of compulsory education, preferring the old apprenticeship system and vocational training. He condemned trade unions. Public indifference to adulteration of food and the popular idea that luxury promotes trade were among the contemporary commonplaces for which he had no sympathy. Progress, he proclaimed, may often be no more than change, and may be change for the worse; new inventions may be turned to good or evil purposes. Like Carlyle, he commended the Germans as the true exponents of the Protestant virtues, and he deplored what he regarded as the paucity of intellectual or moral ideas among his English contemporaries. About these views there will be nearly as many opinions as readers. But it should be the purpose of the essayist not so much to convey information as to stimulate thought, and in this respect the final study is the most challenging of all the essays selected for this volume.

David Ogg

Note: The footnotes in square brackets have been supplied by the present editor

THE SCIENCE OF HISTORY[1]

Ladies and gentlemen,—I have undertaken to speak to you this evening on what is called the Science of History. I fear it is a dry subject; and there seems, indeed, something incongruous in the very connection of such words as Science and History. It is as if we were to talk of the colour of sound, or the longitude of the rule-of-three. Where it is so difficult to make out the truth on the commonest disputed fact in matters passing under our very eyes, how can we talk of a science in things long past, which come to us only through books? It often seems to me as if History was like a child's box of letters, with which we can spell any word we please. We have only to pick out such letters as we want, arrange them as we like, and say nothing about those which do not suit our purpose.

I will try to make the thing intelligible, and I will try not to weary you; but I am doubtful of my success either way. First, however, I wish to say a word or two about the eminent person whose name is connected with this way of looking at History, and whose premature death struck us all with such a sudden sorrow. Many of you, perhaps, recollect Mr Buckle as he stood not so long ago in this place. He spoke for more than an hour without a note—never repeating himself, never wasting words; laying out his matter as easily and as pleasantly as if he had been talking to us at his own fireside. We might think what we pleased of Mr Buckle's views, but it was plain enough that he was a man of uncommon power; and he had qualities also—qualities to which he, perhaps, himself attached little value, as rare as they were admirable.

Most of us, when we have hit on something which we are pleased to think important and original, feel as if we should burst with it. We come out into the book-market with our wares in hand, and ask for thanks and recognition. Mr Buckle, at an early age, conceived the thought which made him famous, but he took the measure of his abilities. He knew that

[1] A lecture delivered at the Royal Institution, February 5, 1864.

whenever he pleased he could command personal distinction, but he cared more for his subject than for himself. He was contented to work with patient reticence, unknown and unheard of, for twenty years; and then, at middle life, he produced a work which was translated at once into French and German, and, of all places in the world, fluttered the dovecotes of the Imperial Academy at St Petersburg.

Goethe says somewhere, that as soon as a man has done anything remarkable, there seems to be a general conspiracy to prevent him from doing it again. He is feasted, fêted, caressed: his time is stolen from him by breakfasts, dinners, societies, idle businesses of a thousand kinds. Mr Buckle had his share of all this; but there are also more dangerous enemies that wait upon success like this. He had scarcely won for himself the place which he deserved, than his health was found shattered by his labours. He had but time to show us how large a man he was—time just to sketch the 'outlines of his philosophy, and he passed away as suddenly as he appeared. He went abroad to recover strength for his work, but his work was done with and over. He died of a fever at Damascus, vexed only that he was compelled to leave it uncompleted. Almost his last conscious words were, ' My book, my book! I shall never finish my book!' He went away as he had lived, nobly careless of himself, and thinking only of the thing which he had undertaken to do.

But his labour had not been thrown away. Disagree with him as we might, the effect which he had already produced was unmistakable, and it is not likely to pass away. What he said was not essentially new. Some such interpretation of human things is as early as the beginning of thought. But Mr Buckle, on the one hand, had the art which belongs to men of genius; he could present his opinions with peculiar distinctness; and, on the other hand, there is much in the mode of speculation at present current among us for which these opinions have an unusual fascination. They do not please us, but they excite and irritate us. We are angry with them; and we betray, in being so, an uneasy misgiving that there may be more truth in those opinions than we like to allow.

Mr Buckle's general theory was something of this kind: When human creatures began first to look about them in the world they lived in, there seemed to be no order in anything. Days and nights were not the same length. The air was sometimes hot and sometimes cold. Some of the stars rose and set like the sun; some were almost motionless in the sky; some described circles round a central star above the north horizon. The planets went on principles of their own; and in the elements there seemed nothing but caprice. Sun and moon would at times go out in eclipse. Sometimes the earth itself would shake under men's feet; and they could only suppose that earth and air and sky and water were inhabited and managed by creatures as wayward as themselves.

Time went on, and the disorder began to arrange itself. Certain influences seemed beneficent to men, others malignant and destructive, and the world was supposed to be animated by good spirits and evil spirits, who were continually fighting against each other, in outward nature and in human creatures themselves. Finally, as men observed more and imagined less, these interpretations gave way also. Phenomena the most opposite in effect were seen to be the result of the same natural law. The fire did not burn the house down if the owners of it were careful, but remained on the hearth and boiled the pot; nor did it seem more inclined to burn a bad man's house down than a good man's, provided the badness did not take the form of negligence. The phenomena of nature were found for the most part to proceed in an orderly, regular way, and their variations to be such as could be counted upon. From observing the order of things, the step was easy to cause and effect. An eclipse, instead of being a sign of the anger of Heaven, was found to be the necessary and innocent result of the relative position of sun, moon, and earth. The comets became bodies in space, unrelated to the beings who had imagined that all creation was watching them and their doings. By degrees, caprice, volition, all symptoms of arbitrary action, disappeared out of the universe; and almost every phenomenon in earth or heaven was found attributable to some law, either understood or perceived to exist. Thus nature was reclaimed from the imagination. The first fantastic conception of things gave way before the moral; the moral in turn gave way before the natural; and at last there was left but one small tract of jungle where the theory of law had failed to penetrate—the doings and characters of human creatures themselves.

There, and only there, amidst the conflicts of reason and emotion, conscience and desire, spiritual forces were still conceived to exist. Cause and effect were not traceable when there was a free volition to disturb the connection. In all other things, from a given set of conditions, the consequences necessarily followed. With man, the word law changed its meaning; and instead of a fixed order, which he could not choose but follow, it became a moral precept, which he might disobey if he dared.

This it was which Mr Buckle disbelieved. The economy which prevailed throughout nature, he thought it very unlikely should admit of this exception. He considered that human beings acted necessarily from the impulse of outward circumstances upon their mental and bodily condition at any given moment. Every man, he said, acted from a motive; and his conduct was determined by the motive which affected him most powerfully. Every man naturally desires what he supposes to be good for him; but to do well, he must know well. He will eat poison, so long as he does not know that it is poison. Let him see that it will kill him and he will not touch it. The question is not of moral right and wrong. Once let him be thoroughly made to feel that the thing is destructive, and he will leave it alone by the law of his nature. His virtues are the result of knowledge; his faults, the necessary consequence of the want of it. A boy desires to draw. He knows nothing about it; he draws men like trees or houses, with their centre of gravity anywhere. He makes mistakes, because he knows no better. We do not blame him. Till he is better taught he cannot help it. But his instruction begins. He arrives at straight lines; then at solids; then at curves. He learns perspective, and light and shade. He observes more accurately the forms which he wishes to represent. He perceives effects, and he perceives the means by which they are produced. He has learned what to do; and, in part, he has learned how to do it; his after-progress will depend on the amount of force which his nature possesses. But all this is as natural as the growth of an acorn. You do not preach to the acorn that it is its duty to become a large tree; you do not preach to the art-pupil that it is his duty to become a Holbein. You plant your acorn in favourable soil, where it can have light and air, and be sheltered from the wind; you remove the superfluous branches, you train the strength into the leading shoots. The acorn will then become as fine a tree as it has vital force to become. The difference between men and other things is only in the largeness and variety of man's capacities; and in this special capacity, that he alone has the power of observing the circumstances favourable to his own growth, and can apply them for himself. Yet, again, with this condition,—that he is not, as is commonly supposed, free to choose whether he will make use of these appliances or not. When he knows what is good for him, he will choose it; and he will judge what is good for him by the circumstances which have made him what he is.

And what he would do, Mr Buckle supposed that he always had done. His history had been a natural growth as much as the growth of the acorn. His improvement had followed the progress of his knowledge; and, by a comparison of his outward circumstances with the condition of his mind, his whole proceedings on this planet, his creeds and constitutions, his good deeds and his bad, his arts and his sciences, his empires and his revolutions, would be found all to arrange themselves into clear relations of cause and effect.

If, when Mr. Buckle pressed his conclusions, we objected the difficulty of finding what the truth about past times really was, he would admit it candidly as far as concerned individuals; but there was not the same difficulty, he said, with masses of men. We might disagree about the characters of Julius or Tiberius Cæsar, but we could know well enough the Romans of the Empire. We had their literature to tell us how they thought; we had their laws to tell us how they governed; we had the broad face of the world, the huge mountainous outline of their general doing upon it, to tell us how they acted. He believed it was all reducible to laws, and could be made as intelligible as the growth of the chalk cliffs or the coal measures.

And thus consistently Mr Buckle cared little for individuals. He did not believe (as some one has said) that the history of mankind is the history of its great men. Great men with him were but larger atoms, obeying the same impulses with the rest, only perhaps a trifle more erratic. With them or without them, the course of things would have been much the same.

As an illustration of the truth of his view, he would point to the new science of Political Economy. Here already was a large area of human activity in which natural laws were found to act unerringly. Men had gone on for centuries trying to regulate trade on moral principles. They had endeavoured to fix wages according to some imaginary rule of fairness; to fix prices by what they considered things ought to cost. They encouraged one trade or discouraged another, for moral reasons. They might as well have tried to work a steam engine on moral reasons. The great statesmen whose names were connected with these enterprises might have as well legislated that water should run uphill. There were natural laws fixed in the conditions of things : and to contend against them was the old battle of the Titans against the gods.

As it was with political economy, so it was with all other forms of human activity; and as the true laws of political economy explained the troubles which people fell into in old times, because they were ignorant of them, so the true laws of human nature, as soon as we knew them, would explain their mistakes in more serious matters, and enable us to manage better for the future. Geographical position, climate, air, soil, and the like, had their several influences. The northern nations are hardy and industrious, because they must till the earth if they would eat the fruits of it, and because the temperature is too low to make an idle life enjoyable. In the south, the soil is more productive, while less food is wanted and fewer clothes; and in the exquisite air, exertion is not needed to make the sense of existence delightful. Therefore, in the south we find men lazy and indolent.

True, there are difficulties in these views; the home of the languid Italian was the home also of the sternest race of whom the story of mankind retains a record. And again, when we are told that the Spaniards are superstitious, because Spain is a country of earthquakes, we remember Japan, the spot in all the world where earthquakes are most frequent, and where at the same time there is the most serene disbelief in any supernatural agency whatsoever.

Moreover, if men grow into what they are by natural laws, they cannot help being what they are, and if they cannot help being what they are, a good deal will have to be altered in our general view of human obligations and responsibilities.

That, however, in these theories there is a great deal of truth is quite certain; were there but a hope that those who maintain them would be contented with this admission. A man born in a Mahometan country grows up a Mahometan; in a Catholic country, a Catholic; in a Protestant country, a Protestant. His opinions are like his language; he learns to think as he learns to speak; and it is absurd to suppose him responsible for being what nature makes him. We take pains to educate children. There is a good education and a bad education; there are rules well ascertained by which characters are influenced, and, clearly enough, it is no mere matter for a boy's free will whether he turns out well or ill. We try to train him into good habits; we keep him out of the way of temptations; we see that he is well taught; we mix kindness and strictness; we surround him with every good influence we can command. These are what are termed the advantages of a good education : and if we fail to provide those under our care with it, and if they go wrong in consequence, the responsibility we feel to be as much ours as theirs. This is at once an admission of the power over us of outward circumstances.

In the same way, we allow for the strength of temptations, and the like.

In general, it is perfectly obvious that men do necessarily absorb, out of the influences in which they grow up, something which gives a complexion of their whole after-character.

When historians have to relate great social or speculative changes, the overthrow of a monarchy or the establishment of a creed, they do but half their duty if they merely relate the events. In an account, for instance, of the rise of Mahometanism, it is not enough to describe the character of the Prophet, the ends which he set before him, the means which he made use of, and the effect which he produced; the historian must show what there was in the condition of the Eastern races which enabled Mahomet to act upon them so powerfully; their existing beliefs, their existing moral and political condition.

In our estimate of the past, and in our calculations of the future—in the judgments which we pass upon one another, we measure responsibility, not by the thing done, but by the opportunities which people have had of knowing better or worse. In the efforts which we make to keep our children from bad associations or friends we admit that external circumstances have a powerful effect in making men what they are.

But are circumstances everything? That is the whole question. A science of history, if it is more than a misleading name, implies that the relation between cause and effect holds in human things as completely as in all others, that the origin of human actions is not to be looked for in mysterious properties of the mind, but in influences which are palpable and ponderable.

When natural causes are liable to be set aside and neutralized by what is called volition, the word Science is out of place. If it is free to a man to choose what he will do or not do, there is no adequate science of him. If there is a science of him, there is no free choice, and the praise or blame with which we regard one another are impertinent and out of place.

I am trespassing upon these ethical grounds because, unless I do, the subject cannot be made intelligible. Mankind are but an aggregate of individuals—History is but the record of individual action; and what is true of the part, is true of the whole.

We feel keenly about such things, and when the logic becomes perplexing, we are apt to grow rhetorical and passionate. But rhetoric is only misleading. Whatever the truth may be, it is best that we should know it; and for truth of any kind we should keep our heads and hearts as cool as we can.

I will say at once, that if we had the whole case before us
—if we were taken, like Leibniz's[2] Tarquin, into the council-
chamber of nature, and were shown what we really were,
where we came from, and where we were going, however
unpleasant it might be for some of us to find ourselves, like
Tarquin, made into villains, from the subtle necessities of
'the best of all possible worlds;' nevertheless, some such
theory as Mr Buckle's might possibly turn out to be true.
Likely enough, there is some great 'equation of the universe'
where the value of the unknown quantities can be determined.
But we must treat things in relation to our own powers and
position; and the question is, whether the sweep of those vast
curves can be measured by the intellect of creatures of a day
like ourselves.

The 'Faust' of Goethe, tired of the barren round of earthly
knowledge, calls magic to his aid. He desires, first, to see the
spirit of the Macrocosmos, but his heart fails him before he
ventures that tremendous experiment, and he summons before
him, instead, the spirit of his own race. There he feels himself
at home. The stream of life and the storm of action, the
everlasting ocean of existence, the web and the woof, and the
roaring loom of time—he gazes upon them all, and in
passionate exultation claims fellowship with the awful thing
before him. But the majestic vision fades, and a voice comes
to him—'Thou art fellow with the spirits which thy mind
can grasp—not with me.'

Had Mr Buckle tried to follow his principles into detail,
it might have fared no better with him than with 'Faust.'

What are the conditions of a science? and when may any
subject be said to enter the scientific stage? I suppose when
the facts of it begin to resolve themselves into groups; when
phenomena are no longer isolated experiences, but appear in
connection and order; when, after certain antecedents, certain
consequences are uniformly seen to follow; when the facts
enough have been collected to furnish a basis for conjectural
explanation, and when conjectures have so far ceased to be

[2] [G. W. Leibniz (1646-1716), German philosopher and mathema-
tician who, in his *Theodicée*, defended freedom of the will.]

utterly vague, that it is possible in some degree to foresee the future by the help of them.

Till a subject has advanced as far as this, to speak of a science of it is an abuse of language. It is not enough to say that there must be a science of human things, because there is a science of all other things. This is like saying the planets must be inhabited, because the only planet of which we have any experience is inhabited. It may or may not be true, but it is not a practical question; it does not affect the practical treatment of the matter in hand.

Let us look at the history of Astronomy.

So long as sun, moon, and planets were supposed to be gods or angels; so long as the sword of Orion was not a metaphor, but a fact, and the groups of stars which inlaid the floor of heaven were the glittering trophies of the loves and wars of the Pantheon, so long there was no science of Astronomy. There was fancy, imagination, poetry, perhaps reverence, but no science. As soon, however, as it was observed that the stars retained their relative places—that the times of their rising and setting varied with the seasons—that sun, moon, and planets moved among them in a plane, and the belt of the Zodiac was marked out and divided, then a new order of things began. Traces of the earlier stage remained in the names of the signs and constellations, just as the Scandinavian mythology survives now in the names of the days of the week: but for all that, the understanding was now at work on the thing; Science had begun, and the first triumph of it was the power of foretelling the future. Eclipses were perceived to recur in cycles of nineteen years, and philosophers were able to say when an eclipse was to be looked for. The periods of the planets were determined. Theories were invented to account for their eccentricities; and, false as those theories might be, the position of the planets could be calculated with moderate certainty by them. The very first result of the science, in its most imperfect stage, was a power of foresight; and this was possible before any one true astronomical law had been discovered.

We should not therefore question the possibility of a science of history, because the explanations of its phenomena were rudimentary or imperfect: that they might be, and might long continue to be, and yet enough might be done to show that there was such a thing, and that it was not entirely without use. But how was it that in those rude days, with small knowledge of mathematics, and with no better instruments than flat walls and dial plates, the first astronomers made progress so considerable? Because, I suppose, the phenomena which they were observing recurred, for the most part, within moderate intervals; so that they could collect large experience within the compass of their natural lives: because days and months and years were measurable periods, and within them the more simple phenomena perpetually repeated themselves.

But how would it have been if, instead of turning on its axis in twenty-four hours, the earth had taken a year about it; if the year had been nearly four-hundred years; if man's life had been no longer than it is, and for the initial steps of astronomy there had been nothing to depend upon except observations recorded in history? How many ages would have passed, had this been our condition, before it would have occurred to any one, that, in what they saw night after night, there was any kind of order at all?

We can see to some extent how it would have been, by the present state of those parts of the science which in fact depend on remote recorded observations. The movements of the comets are still extremely uncertain. The times of their return can be calculated only with the greatest vagueness.

And yet such a hypothesis as I have suggested would but inadequately express the position in which we are in fact placed towards history. There the phenomena never repeat themselves. There we are dependent wholly on the record of things said to have happened once, but which never happen or can happen a second time. There no experiment is possible; we can watch for no recurring fact to test the worth of our conjectures. It has been suggested, fancifully, that if we consider the universe to be infinite, time is the same as eternity, and the past is perpetually present. Light takes nine years to come to us from Sirius; those rays which we may see to-night when we leave this place, left Sirius nine years ago; and could the inhabitants of Sirius see the earth at this moment, they would see the English army in the trenches before Sebastopol; Florence Nightingale watching at Scutari over the wounded at Inkermann; and the peace of England undisturbed by ' Essays and Reviews.'[3]

As the stars recede into distance, so time recedes with them, and there may be, and probably are, stars from which Noah might be seen stepping into the ark, Eve listening to the temptation of the serpent, or that older race, eating the oysters and leaving the shell-heaps behind them, when the Baltic was an open sea.

Could we but compare notes, something might be done; but of this there is no present hope, and without it there will be no science of history. Eclipses, recorded in ancient books, can be verified by calculation, and lost dates can be recovered by them, and we can foresee by the laws which they follow when there will be eclipses again. Will a time ever be when the lost secret of the foundation of Rome can be recovered by historic laws? If not, where is our science? It may be said that this is a particular fact, that we can deal satisfactorily with general phenomena affecting eras and cycles. Well, then, let us take some general phenomenon. Mahometanism, for instance, or Buddhism. Those are large enough. Can you imagine a science which would have[4] *foretold* such movements as those? The state of things out of which they rose is obscure; but suppose it not obscure, can you conceive that, with any amount of historical insight into the old Oriental beliefs, you could have seen that they were about to transform themselves into those particular forms and no other?

It is not enough to say, that, after the fact, you can understand partially how Mahometanism came to be. All historians worth the name have told us something about that. But when we talk of science, we mean something with more ambitious

---

[3] [A volume of essays on theological subjects published in 1860. Some of the contributors were prosecuted for heresy.]

[4] It is objected that Geology is a science: yet that Geology cannot foretell the future changes of the earth's surface. Geology is not a century old, and its periods are measured by millions of years. Yet, if Geology cannot foretell future facts, it enabled Sir Roderick Murchison to foretell the discovery of Australian gold.

pretences, we mean something which can foresee as well as explain; and, thus looked at, to state the problem is to show its absurdity. As little could the wisest man have foreseen this mighty revolution, as thirty years ago such a thing as Mormonism could have been anticipated in America; as little as it could have been foreseen that table-turning and spirit-rapping would have been an outcome of the scientific culture of England in the nineteenth century.

The greatest of Roman thinkers gazing mournfully at the seething mass of moral putrefaction round him, detected and deigned to notice among its elements a certain detestable superstition, so he called it, rising up amidst the offscouring of the Jews, which was named Christianity. Could Tacitus have looked forward nine centuries to the Rome of Gregory VII, could he have beheld the representative of the majesty of the Cæsars holding the stirrup of the Pontiff of that vile and execrated sect, the spectacle would scarcely have appeared to him the fulfilment of a rational expectation, or an intelligible result of the causes in operation round him. Tacitus, indeed, was born before the science of history; but would M. Comte[5] have seen any more clearly?

Nor is the case much better if we are less hard upon our philosophy; if we content ourselves with the past, and require only a scientific explanation of that.

First, for the facts themselves. They come to us through the minds of those who recorded them, neither machines nor angels, but fallible creatures, with human passions and prejudices. Tacitus and Thucydides were perhaps the ablest men who ever gave themselves to writing history; the ablest and also the most incapable of conscious falsehood. Yet even now, after all these centuries, the truth of what they relate is called in question. Good reasons can be given to show that neither of them can be confidently trusted. If we doubt with these, whom are we to believe?

Or again, let the facts be granted. To revert to my simile of the box of letters, you have but to select such facts as suit you,

[5] [Auguste Comte (1798-1857), French philosopher who popularised the ' Positivist Philosophy '.]

you have but to leave alone those which do not suit you, and let your theory of history be what it will, you can find no difficulty in providing facts to prove it.

You may have your Hegel's philosophy of history, or you may have your Schlegel's philosophy of history; you may prove from history that the world is governed in detail by a special Providence; you may prove that there is no sign of any moral agent in the universe, except man; you may believe, if you like it, in the old theory of the wisdom of antiquity; you may speak, as was the fashion in the fifteenth century, of ' our fathers, who had more wit and wisdom than we;' or you may talk of ' our barbarian ancestors,' and describe their wars as the scuffling of kites and crows.

You may maintain that the evolution of humanity has been an unbroken progress towards perfection; you may maintain that there has been no progress at all, and that man remains the same poor creature that he ever was; or, lastly, you may say with the author of the ' Contrat Social,' that men were purest and best in primeval simplicity—

When wild in woods the noble savage ran.

In all, or any of these views, history will stand your friend. History, in its passive irony, will make no objection. Like Jarno, in Goethe's novel, it will not condescend to argue with you, and will provide you with abundant illustrations of anything which you may wish to believe.

' What is history,' said Napoleon, ' but a fiction agreed upon?' ' My friend,' said Faust to the student, who was growing enthusiastic about the spirit of past ages; ' my friend, the times which are gone are a book with seven seals, and what you call the spirit of past ages is but the spirit of this or that worthy gentleman in whose mind those ages are reflected.'

One lesson, and only one, history may be said to repeat with distinctness; that the world is built somehow on moral foundations; that, in the long run, it is well with the good; in the long run, it is ill with the wicked. But this is no science; it is no more than the old doctrine taught long ago by the Hebrew prophets. The theories of M. Comte and his disciples advance us, after all, not a step beyond the trodden and familiar ground. If men are not entirely animals, they are at least half animals, and are subject in this aspect of them to the conditions of animals. So far as those parts of man's doings are concerned, which neither have, nor need have, anything moral about them, so far the laws of him are calculable. There are laws for his digestion, and laws of the means by which his digestive organs are supplied with matter. But pass beyond them, and where are we? In a world where it would be as easy to calculate men's actions by laws like those of positive philosophy as to measure the orbit of Neptune with a foot-rule, or weigh Sirius in a grocer's scale.

And it is not difficult to see why this should be. The first principle on which the theory of a science of history can be plausibly argued, is that all actions whatsoever arise from self-interest. It may be enlightened self-interest; it may be unenlightened; but it is assumed as an axiom, that every man, in whatever he does, is aiming at something which he considers will promote his happiness. His conduct is not determined by his will; it is determined by the object of his desire. Adam Smith, in laying the foundation of political economy, expressly eliminates every other motive. He does not say that men never act on other motives; still less, that they never ought to act on other motives. He asserts merely that, as far as the arts of production are concerned, and of buying and selling, the action of self-interest may be counted upon as uniform. What Adam Smith says of political economy, Mr Buckle would extend over the whole circle of human activity.

Now, that which especially distinguishes a high order of man from a low order of man—that which constitutes human goodness, human greatness, human nobleness—is surely not the degree of enlightenment with which men pursue their own advantage; but it is self-forgetfulness—it is self-sacrifice—it is the disregard of personal pleasure, personal indulgence, personal advantages remote or present, because some other line of conduct is more right.

We are sometimes told that this is but another way of expressing the same thing; that when a man prefers doing what is right, it is only because to do right gives him a higher satisfaction. It appears to me, on the contrary, to be a difference in the very heart and nature of things. The martyr goes to the stake, the patriot to the scaffold, not with a view to any future reward to themselves, but because it is a glory to fling away their lives for truth and freedom. And so through all phases of existence, to the smallest details of common life, the beautiful character is the unselfish character. Those whom we most love and admire are those to whom the thought of self seems never to occur; who do simply and with no ulterior aim—with no thought whether it will be pleasant to themselves or unpleasant—that which is good, and right, and generous.

Is this still selfishness, only more enlightened? I do not think so. The essence of true nobility is neglect of self. Let the thought of self pass in, and the beauty of a great action is gone—like the bloom from a soiled flower. Surely it is a paradox to speak of the self-interest of a martyr who dies for a cause, the triumph of which he will never enjoy; and the greatest of that great company in all ages would have done what they did, had their personal prospects closed with the grave. Nay, there have been those so zealous for some glorious principle, as to wish themselves blotted out of the book of Heaven if the cause of Heaven could succeed.

And out of this mysterious quality, whatever it be, arise the higher relations of human life, the higher modes of human obligation. Kant, the philosopher, used to say that there were two things which overwhelmed him with awe as he thought of them. One was the star-sown deep of space, without limit and without end; the other was, right and wrong. Right, the sacrifice of self to good; wrong, the sacrifice of good to self; —not graduated objects of desire, to which we are determined by the degrees of our knowledge, but wide asunder as pole and pole, as light and darkness—one, the object of infinite love; the other, the object of infinite detestation and scorn. It is in this marvellous power in men to do wrong (it is an old story, but none the less true for that)—it is in this power to do wrong—wrong or right, as it lies somehow with ourselves to choose—that the impossibility stands of forming scientific calculations of what men will do before the fact, or scientific explanations of what they have done after the fact. If men were consistently selfish, you might analyze their motives; if they were consistently noble, they would express in their conduct the laws of the highest perfection. But so long as two natures are mixed together, and the strange creature which results from the combination is now under one influence and now under another, so long as you will make nothing of him except from the old-fashioned moral—or, if you please, imaginative—point of view.

Even the laws of political economy itself cease to guide us when they touch moral government. So long as labour is a chattel to be bought and sold, so long, like other commodities, it follows the condition of supply and demand. But if, for his misfortune, an employer considers that he stands in human relations towards his workmen; if he believes, rightly or wrongly, that he is responsible for them; that in return for their labour he is bound to see that their children are decently taught, and they and their families decently fed, and clothed, and lodged; that he ought to care for them in sickness and in old age; then political economy will no longer direct him, and the relations between himself and his dependents will have to be arranged on other principles.

So long as he considers only his own material profit, so long supply and demand will settle every difficulty: but the introduction of a new factor spoils the equation.

And it is precisely in this debatable ground of low motives and noble emotions—in the struggle, ever failing, yet ever renewed, to carry truth and justice into the administration of human society; in the establishment of states and in the overthrow of tyrannies; in the rise and fall of creeds; in the world of ideas; in the character and deeds of the great actors in the drama of life; where good and evil fight out their everlasting battle, now ranged in opposite camps, now and more often in the heart, both of them, of each living man—that the true human interest of history resides. The progress of industries, the growth of material and mechanical civilization, are interesting, but they are not the most interesting. They have their reward in the increase of material comforts; but, unless we are mistaken about our nature, they do not highly concern us after all.

Once more; not only is there in men this baffling duality of principle, but there is something else in us which still more defies scientific analysis.

Mr Buckle would deliver himself from the eccentricities of this and that individual by a doctrine of averages. Though he cannot tell whether A, B, or C will cut his throat, he may assure himself that one man in every fifty thousand, or thereabout (I forget the exact proportion), will cut his throat, and with this he consoles himself. No doubt it is a comforting discovery. Unfortunately, the average of one generation need not be the average of the next. We may be converted by the Japanese, for all that we know, and the Japanese methods of taking leave of life may become fashionable among us. Nay, did not Novalis suggest that the whole race of men would at last become so disgusted with their impotence, that they would extinguish themselves by a simultaneous act of suicide, and make room for a better order of beings? Anyhow, the fountain out of which the race is flowing perpetually changes—no two generations are alike. Whether there is a change in the organization itself, we cannot tell; but this is certain, that as the planet varies with the physical atmosphere which surrounds it, so each new generation varies with the last, because it inhales as its spiritual atmosphere the accumulated experience and knowledge of the whole past of the world. These things form the intellectual air which we breathe as we grow; and in the infinite multiplicity of elements of which that air is now composed, it is for ever matter of conjecture what the minds will be like which expand under its influence.

From the England of Fielding and Richardson to the England of Miss Austen—from the England of Miss Austen to the England of Railways and Freetrade, how vast the change; yet perhaps Sir Charles Grandison[6] would not seem so strange to us now, as one of ourselves will seem to our great-grand-children. The world moves faster and faster; and the difference will probably be considerably greater.

The temper of each new generation is a continual surprise.

[6] [Hero of Samuel Richardson's *History of Sir Charles Grandison* (1753), a novel designed to portray the perfect gentleman.]

The fates delight to contradict our most confident expectations. Gibbon believed that the era of conquerors was at an end. Had he lived out the full life of man, he would have seen Europe at the feet of Napoleon. But a few years ago we believed the world had grown too civilized for war, and the Crystal Palace in Hyde Park was to be the inauguration of a new era. Battles, bloody as Napoleon's, are now the familiar tale of every day; and the arts which have made greatest progress are the arts of destruction. What next? We may strain our eyes into the future which lies beyond this waning century; but never was conjecture more at fault. It is blank darkness, which even the imagination fails to people.

What then is the use of History? and what are its lessons? If it can tell us little of the past, and nothing of the future, why waste our time over so barren a study?

First, it is a voice for ever sounding across the centuries the laws of right and wrong. Opinions alter, manners change, creeds rise and fall, but the moral law is written on the tablets of eternity. For every false word or unrighteous deed, for cruelty and oppression, for lust or vanity, the price has to be paid at last : not always by the chief offenders, but paid by some one. Justice and truth alone endure and live. Injustice and falsehood may be long-lived, but doomsday comes at last to them, in French revolutions and other terrible ways.

That is one lesson of History. Another is, that we should draw no horoscopes; that we should expect little, for what we expect will not come to pass. Revolutions, reformations—those vast movements into which heroes and saints have flung themselves, in the belief that they were the dawn of the millennium—have not borne the fruit which they looked for. Millenniums are still far away. These great convulsions leave the world changed, perhaps improved,—but not improved as the actors in them hoped it would be. Luther would have gone to work with less heart, could he have foreseen the Thirty Years' War, and in the distance the theology of Tubingen.[7] Washington might have hesitated to draw the sword against

[7] [Famous university in Wurttemberg, noted for the advanced views of its theological professors.]

England, could he have seen the country which he made as we see it now.[8]

The most reasonable anticipations fail us—antecedents the most opposite mislead us; because the conditions of human problems never repeat themselves. Some new feature alters everything—some element which we detect only in its after-operation.

But this, it may be said, is but a meagre outcome. Can the long records of humanity, with all its joys and sorrows, its sufferings and its conquests, teach us no more than this? Let us approach the subject from another side.

If you were asked to point out the special features in which Shakespeare's plays are so transcendently excellent, you would mention, perhaps, among others, this, that his stories are not put together, and his characters are not conceived, to illustrate any particular law or principle. They teach many lessons, but not any one prominent above another; and when we have drawn from them all the direct instruction which they contain, there remains still something unresolved— something which the artist gives, and which the philosopher cannot give.

It is in this characteristic that we are accustomed to say Shapespeare's supreme *truth* lies. He represents real life. His dramas teach as life teaches—neither less nor more. He builds his fabrics as nature does, on right and wrong; but he does not struggle to make nature more systematic than she is. In the subtle interflow of good and evil—in the unmerited sufferings of innocence—in the disproportion of penalties to desert—in the seeming blindness with which justice, in attempting to assert itself, overwhelms innocent and guilty in a common ruin—Shakespeare is true to real experience. The mystery of life he leaves as he finds it; and, in his most tremendous positions, he is addressing rather the intellectual emotions than the understanding,—knowing well that the understanding in such things is at fault, and the sage as ignorant as the child.

Only the highest order of genius can represent nature thus. An inferior artist produces either something entirely immoral, where good and evil are names, and nobility of disposition

[8] February, 1864.

is supposed to show itself in the absolute disregard of them—or else, if he is a better kind of man, he will force on nature a didactic purpose; he composes what are called moral tales, which may edify the conscience, but only mislead the intellect.

The finest work of this kind produced in modern times is Lessing's play of ' Nathan the Wise.' The object of it is to teach religious toleration. The doctrine is admirable—the mode in which it is enforced is interesting; but it has the fatal fault, that it is not true. Nature does not teach religious toleration by any such direct method; and the result is—no one knew it better than Lessing himself—that the play is not poetry, but only splendid manufacture. Shakespeare is eternal; Lessing's ' Nathan ' will pass away with the mode of thought which gave it birth. One is based on fact; the other, on human theory about fact. The theory seems at first sight to contain the most immediate instruction; but it is not really so.

Cibber and others, as you know, wanted to alter Shakespeare. The French King, in ' Lear,' was to be got rid of; Cordelia was to marry Edgar, and Lear himself was to be rewarded for his sufferings by a golden old age. They could not bear that Hamlet should suffer for the sins of Claudius. The wicked king was to die, and the wicked mother; and Hamlet and Ophelia were to make a match of it, and live happily ever after. A common novelist would have arranged it thus; and you would have had your comfortable moral that wickedness was fitly punished, and virtue had its due reward, and all would have been well. But Shakespeare would not have it so. Shakespeare knew that crime was not so simple in its consequences, or Providence so paternal. He was contented to take the truth from life; and the effect upon the mind of the most correct theory of what life ought to be, compared to the effect of the life itself, is infinitesimal in comparison.

Again, let us compare the popular historical treatment of remarkable incidents with Shakespeare's treatment of them. Look at ' Macbeth.' You may derive abundant instruction from it—instruction of many kinds. There is a moral lesson of profound interest in the steps by which a noble nature glides to perdition. In more modern fashion you may speculate, if you like, on the political conditions represented there, and the temptation presented in absolute monarchies to unscrupulous ambition; you may say, like Dr. Slop,[9] these things could not have happened under a constitutional government; or, again, you may take up your parable against superstition—you may dilate on the frightful consequences of a belief in witches, and reflect on the superior advantages of an age of schools and newspapers. If the bare facts of the story had come down to us from a chronicler, and an ordinary writer of the nineteenth century had undertaken to relate them, his account, we may depend upon it, would have been put together upon one or other of these principles. Yet, by the side of that unfolding of the secrets of the prison-house of the soul, what lean and shrivelled anatomies the best of such descriptions would seem!

Shakespeare himself, I suppose, could not have given us a theory of what he meant—he gave us the thing itself, on which we might make whatever theories we pleased.

Or again, look at Homer.

The ' Iliad ' is from two to three thousand years older than ' Macbeth,' and yet it is as fresh as if it had been written yesterday. We have there no lessons save in the emotions which rise in us as we read. Homer had no philosophy; he never struggles to impress upon us his views about this or that; you can scarcely tell indeed whether his sympathies are Greek or Trojan; but he represents to us faithfully the men and women among whom he lived. He sang the Tale of Troy, he touched his lyre, he drained the golden beaker in the halls of men like those on whom he was conferring immortality. And thus, although no Agamemnon, king of men, ever led a Grecian fleet to Ilium; though no Priam sought the midnight tent of Achilles; though Ulysses and Diomed and Nestor were but names, and Helen but a dream, yet, through Homer's power of representing men and women, those old Greeks will still stand out from amidst the darkness of the ancient world with a sharpness of outline which belongs to no period of history except the most recent. For the mere hard purposes of history, the ' Iliad ' and ' Odyssey ' are the most effective books which ever were written. We see the

9 [A character in Laurence Sterne's *Tristam Shandy* (1760-1768).]

Hall of Menelaus, we see the garden of Alcinous, we see Nausicaa among her maidens on the shore, we see the yellow monarch sitting with ivory sceptre in the Market-place dealing out genial justice. Or again, when the wild mood is on, we can hear the crash of the spears, the rattle of the armour as the heroes fall, and the plunging of the horses among the slain. Could we enter the palace of an old Ionian lord, we know what we should see there; we know the words in which he would address us. We could meet Hector as a friend. If we could choose a companion to spend an evening with over a fireside, it would be the man of many counsels, the husband of Penelope.

I am not going into the vexed question whether History or Poetry is the more true. It has been sometimes said that Poetry is the more true, because it can make things more like what our moral sense would prefer they should be. We hear of poetic justice and the like, as if nature and fact were not just enough.

I entirely dissent from that view. So far as poetry attempts to improve on truth in that way, so far it abandons truth, and is false to itself. Even literal facts, exactly as they were, a great poet will prefer whenever he can get them. Shakespeare in the historical plays is studious, wherever possible, to give the very words which he finds to have been used; and it shows how wisely he was guided in this, that those magnificent speeches of Wolsey are taken exactly, with no more change than the metre makes necessary, from Cavendish's Life. Marlborough read Shakespeare for English history, and read nothing else. The poet only is not bound, when it is inconvenient, to what may be called the accidents of facts. It was enough for Shakespeare to know that Prince Hal in his youth had lived among loose companions, and the tavern in Eastcheap came in to fill out his picture; although Mrs Quickly and Falstaff, and Poins and Bardolph, were more likely to have been fallen in with by Shakespeare himself at the Mermaid, than to have been comrades of the true Prince Henry. It was enough for Shakespeare to draw real men, and the situation, whatever it might be, would sit easy on them. In this sense only it is that Poetry is truer than History, that it can make a picture more complete. It may take liberties with time and space, and give the action distinctness by throwing it into more manageable compass.

But it may not alter the real conditions of things, or represent life as other than it is. The greatness of the poet depends on his being true to nature, without insisting that nature should theorize with him, without making her more just, more philosophical, more moral than reality; and, in difficult matters, leaving much to reflection which cannot be explained.

And if this be true of Poetry—if Homer and Shakespeare are what they are, from the absence of everything didactic about them—may we not thus learn something of what History should be, and in what sense it should aspire to teach?

If Poetry must not theorize, much less should the historian theorize, whose obligations to be true to fact are even greater than the poet's. If the drama is grandest when the action is least explicable by laws, because then it best resembles life, then history will be grandest also under the same conditions. 'Macbeth,' were it literally true, would be perfect history; and so far as the historian can approach to that kind of model, so far as he can let his story tell itself in the deeds and words of those who act it out, so far is he most successful. His work is no longer the vapour of his own brain, which a breath will scatter; it is the thing itself, which will have interest for all time. A thousand theories may be formed about it— spiritual theories, Pantheistic theories, cause and effect theories; but each age will have its own philosophy of history, and all these in turn will fail and die. Hegel falls out of date, Schlegel falls out of date, and Comte in good time will fall out of date; the thought about the thing must change as we change; but the thing itself can never change; and a history is durable or perishable as it contains more or least of the writer's own speculations. The splendid intellect of Gibbon for the most part kept him true to the right course in this; yet the philosophical chapters for which he has been most admired or censured may hereafter be thought the least interesting in his work. The time has been when they would not have been comprehended: the time may come when they will seem commonplace.

It may be said, that in requiring history to be written like a drama, we require an impossibility.

For history to be written with the complete form of a drama, doubtless is impossible; but there are periods, and these the periods, for the most part, of greatest interest to mankind, the history of which may be so written that the actors shall reveal their characters in their own words; where mind can be seen matched against mind, and the great passions of the epoch not simply be described as existing, but be exhibited at their white heat in the souls and hearts possessed by them. There are all the elements of drama—drama of the highest order—where the huge forces of the times are as the Grecian destiny, and the power of the man is seen either stemming the stream till it overwhelms him, or ruling while he seems to yield to it.

It is Nature's drama—not Shakespeare's—but a drama none the less.

So at least it seems to me. Wherever possible, let us not be told *about* this man or that. Let us hear the man himself speak, let us see him act, and let us be left to form our own opinions about him. The historian, we are told, must not leave his readers to themselves. He must not only lay the facts before them—he must tell them what he himself thinks about those facts. In my opinion, this is precisely what he ought not to do. Bishop Butler says somewhere, that the best book which could be written would be a book consisting only of premises, from which the readers should draw conclusions for themselves. The highest poetry is the very thing which Butler requires, and the highest history ought to be. We should no more ask for a theory of this or that period of history, than we should ask for a theory of ' Macbeth ' or ' Hamlet.' Philosophies of history, sciences of history—all these, there will continue to be; the fashions of them will change, as our habits of thought will change; and each new philosopher will find his chief employment in showing that before him no one understood anything. But the drama of history is imperishable, and the lessons of it will be like what we learn from Homer or Shakespeare—lessons for which we have no words.

The address of history is less to the understanding than to

the higher emotions. We learn in it to sympathize with what is great and good; we learn to hate what is base. In the anomalies of fortune we feel the mystery of our mortal existence, and in the companionship of the illustrious natures who have shaped the fortunes of the world, we escape from the littlenesses which cling to the round of common life, and our minds are tuned in a higher and nobler key.

For the rest, and for those large questions which I touched in connection with Mr Buckle, we live in times of disintegration, and none can tell what will be after us. What opinions—what convictions—the infant of to-day will find prevailing on the earth, if he and it live out together to the middle of another century, only a very bold man would undertake to conjecture! ' The time will come,' said Lichtenberg,[10] in scorn at the materializing tendencies of modern thought; ' the time will come when the belief in God will be as the tales with which old women frighten children; when the world will be a machine, the ether a gas, and God will be a force.' Mankind, if they last long enough on the earth, may develope strange things out of themselves; and the growth of what is called the Positive Philosophy is a curious commentary on Lichtenberg's prophecy. But whether the end be seventy years hence, or seven hundred—be the close of the mortal history of humanity as far distant in the future as its shadowy beginnings seem now to lie behind us—this only we may foretell with confidence—that the riddle of man's nature will remain unsolved. There will be that in him yet which physical laws fail to explain—that something, whatever it be, in himself, and in the world, which science cannot fathom, and which suggests the unknown possibilities of his origin and his destiny. There will remain yet

> *Those obstinate questionings*
> *Of sense and outward things;*
> *Falling from us, vanishings—*
> *Blank misgivings of a creature*
> *Moving about in worlds not realized—*

[10] [Georg Christoph Lichtenberg (1742-1799), German chemist and physicist.]

High instincts, before which our mortal nature
Doth tremble like a guilty thing surprised.
There will remain
    Those first affections—
    Those shadowy recollections—
    Which, be they what they may,
    Are yet the fountain-light of all our day—
    Are yet the master-light of all our seeing—
    Uphold us, cherish, and have power to make
    Our noisy years seem moments in the being
        Of the Eternal Silence.

SOCIETY IN ITALY IN THE LAST DAYS OF THE ROMAN REPUBLIC

Whether free institutions create good citizens, or whether conversely free institutions imply good citizens and wither up and perish as private virtue decays, is a question which will continue to be agitated as long as political society continues. The science of history ought to answer it, but the science of history is silent or ambiguous where, if it could tell us anything at all, it would be able to speak decidedly. What is called the philosophy of history is, and can be, only an attempted interpretation of earlier ages by the modes of thought current in our own; and those modes of thought, being formed by the study of the phenomena which are actually round us, are changed from era to era. We read the past by the light of the present, and the forms vary as the shadows fall, or as the point of vision alters. Those who have studied most conscientiously the influences which have determined their own convictions will be the last to claim exemption from the control of forces which they recognise as universal and irresistible. The foreground of human life is the only part of it which we can examine with real exactness. As the distance recedes details disappear in the shade, or resolve themselves into outlines. We turn to contemporary books and records, but we lose in light and in connection with present experience what we gain in minuteness. The accounts of their own times which earlier writers leave to us are coloured in turn by their opinions, and we cannot so reproduce the past as to guard against prejudices which governed those writers so much as they govern ourselves. The result, even to the keenest historical sight, is no more than a picture which each of us paints for himself upon the retina of his own imagination.

These conditions of our nature warn us all, if we are wise, against generalised views of history. We form general views. This, too, we cannot help, unless we are ignorant of the past altogether. But we receive them for what they are worth. They do not repose upon a knowledge of facts which can form the foundations of a science. We see certain objects; but we see them not as they were, but fore-shortened by distance and coloured by the atmosphere of time. The impression, before it arrives in our minds, has been half created by ourselves. Therefore it is that from philosophy of history, from attempts to explain the phenomena of earlier generations by referring them to general principles, we turn with weariness and distrust. We find more interest in taking advantage of those rare occasions where we can apply a telescope to particular incidents, and catch a sight of small fractions of the actual doings of our fellow-mortals, where accident enables us to examine them in detailed pattern. We may obtain little in this way to convince our judgment, but we can satisfy an innocent curiosity, and we can sometimes see enough to put us on our guard against universal conclusions.

We know, for instance (so far as we can speak of knowledge of the general character of an epoch), that the early commonwealth of Rome was distinguished by remarkable purity of manners; that the marriage tie was singularly respected; that the Latin yeomen, who were the back-bone of the community, were industrious and laborious, that they lived with frugality and simplicity, and brought up their children in a humble fear of God or of the gods as rulers to whom they would one day have to give an account. That the youth of a plant which grew sturdily was exceptionally healthy is no more than we should naturally infer, and that the fact was so is confirmed to us both by legend and authentic record. The change of manners is assumed by some persons to have come in with the Cæsars. Virtue is supposed to have flourished so long as liberty survived, and the perfidy and profligacy of which we read with disgust in Tacitus and Juvenal are regarded as the offspring of despotism. With the general state of European morals under the first centuries of the Empire we are extremely ill-acquainted. Tacitus and Juvenal describe the society of the capital. Of life in the country and in the provincial towns they tell us next to nothing. If we may presume that the Messalinas had their imitators in the provinces; if we may gather from the Epistles of St Paul that the morals of Corinth for instance were not distinguished by any special excellence, yet there was virtue or desire of virtue enough in the world to make possible the growth of Christianity.

Accident, on the other hand, has preserved the fragments of a drama of real life, which was played out in the last days of the Republic, partly in Rome itself, partly in a provincial city in South Italy, from which it would appear that the ancient manners were already everywhere on the decline; that institutions suited to an age when men were a law to themselves, could not prevent them from becoming wicked if they were inclined, and only saved them from punishment when they had deserved it. The broken pieces of the story leave much to be desired. The actions are preserved; the actors are little more than names. The flesh and blood, the thoughts that wrought in the brain, the passions that boiled in the veins—these are dry as the dust of a mummy from an Egyptian catacomb. Though generations pass away, however, the earth at least remains. We cannot see the old nations, but we can stand where they stood; we can look on the landscape on which they looked; we can watch the shadows of the clouds chasing one another on the same mountain slopes; we can listen to the everlasting music of the same water-falls; we can hear the same surf far off breaking upon the beach.

Let us transport ourselves then to the Neopolitan town of Larino, not far from the Gulf of Venice.[1] In the remains of the amphitheatre we can recognise the Roman hands that once were labouring there.

Let us imagine that it is the year 88 before Christ, when Cæsar was a boy of twelve, when the Social War had just been ended by Sylla, and Marius had fled from Rome, to moralise amidst the ruins of Carthage. Larino, like most of the Samnite towns, had taken part with the patriots. Several of its most distinguished citizens had fallen in battle. They had been defeated, but their cause had survived. Summoned to Asia to oppose Mithridates, Sylla had postponed his revenge, and had

[1] [Actually Larino is about 20 miles south of the Adriatic port of Termoli and is a long way from the Gulf of Venice.]

conceded at least some of the objects for which the Italians had been in arms. The leaders returned to their homes, and their estates escaped confiscation. The two families of highest consequence in Larino were the Cluentii and the Aurii. Both were in mourning. Lucius Cluentius, who had commanded the insurgent army in Campania, had been killed at Nola. Marcus Aurius had not returned to Larino at the peace, and was supposed to have fallen in the North of Italy. Common political sympathies had drawn the survivors together, and they were further connected by marriage. There remained of the Cluentii a widowed mother named Sassia, with two children, Aulus Cluentius Avitus, a boy of sixteen, and his sister Cluentia, a year younger. Dinea, the mother of the Aurii, was a widow also. Dinea had been the sister of Sassia's husband, and was therefore herself a Cluentia. She had four children, all some years older than their cousins—Marcus Aurius, whom she believed to be dead; Numerius Aurius, Cnæus Magius Aurius; and a daughter, Magia.

The Aurii had other relations of the same name at Larino—Aurius Melinus, Caius Melinus, and several more. The Cluentii were the last of their race. Both families were rich. The wealth which had poured into Rome after the conquest of the East had filtered over Italy. These provincial magistrates lived in handsome villas, with comforts which would have made Cato shudder, and waited upon by retinues of slaves. Otherwise scandal had no harm to say of either Aurii or Cluentii. They were honoured for their patriotism, and beloved for their private virtues.

A third family at Larino, the Oppianici, though also connected with the Aurii, belonged to the opposite faction. Caius Oppianicus, the younger of two brothers was married to Dinea's daughter Magia. Statius Albinus Oppianicus, the elder, and the head of the clan, had been three times married : first to a sister of Dinea, who had died, leaving him with a son; next, to a lady named Papia, who bore him a son also, and whom he had divorced; lastly, to Novia, who was for the present living with him and had brought him a third son, an infant. He had squandered his own fortune and the fortune of his first wife, whom he was suspected of having poisoned. He had since been living by his wits, and had figured unpleasantly in a late trial at Rome. A foolish youth of Larino, appropriately named Asinius, had come into possession of a large sum of money. Like Iago, who made his fool his purse, Oppianicus took possession of Asinius, carried him to Rome to see the world, and launched him among the taverns and the gambling houses. A confederate, Avilius, a Larinate also, made a third in the party; and one night, when Asinius was absent with a female companion, with whom they were assured that he would remain till morning, Avilius affected to be taken suddenly ill, and said that he must make his will. A notary and witnesses were introduced to whom the persons of Avilius and Asinius were alike unknown. Avilius bequeathed all his property to Oppianicus, signed his name Asinius, and then recovered. The true Asinius was waylaid and killed a few days after. Oppianicus produced the will, claimed the estate, and obtained it—not, however, without some notice having been drawn to the matter which might have ended unpleasantly for him. Suspicions had been aroused, it does not apear how. Avilius was arrested and carried before one of the city magistrates, to whom in his terror he confessed the truth. Fortunately for Oppianicus, the magistrate was discreet and not inaccessible. The spoils were divided and the affair hushed up, but it had naturally been much talked of at Larino. Oppianicus had been looked on askance; in the matter of fortune he was in a desperate condition, and he was on the look-out for the nearest means of improving his circumstances.

He was a man, it appears, of considerable personal attractions. He had made himself agreeable to his brother's wife Magia, and had seduced her. Her brother Numerius caught a fever and suddenly died, leaving his share of the Aurian property to his brother Cnæus Magius.

Cnæus Magius fell ill also very soon after. He, perhaps, suspected the cause of his sickness. At any rate he had seen with alarm and suspicion his sister's intimacy with a person of so questionable a character as Albinus Oppianicus. His alarms were not diminished when her husband, Caius Oppianicus, was found dead in his bed, from some unexplained visitation; and growing rapidly worse, and feeling that his own end was not far off, he sent for his sister, and in the presence of his mother Dinea he questioned her as to whether she was pregnant. She assured him that it was so. She half satisfied him that she was herself innocent of guilt, and that Caius Oppianicus, and not his brother, was the father. He made a will bequeathing the whole inheritance which had fallen to him to this child as soon as it should be born. He appointed his mother, Dinea, the guardian, lest Albinus Oppianicus should interfere. If the child should miscarry, or should not survive, Dinea and Magia were then to divide the estates between them.

The arrangement had scarcely been completed when Cnæus Magius died. Oppianicus then induced Magia to take a medicine which produced abortion. Magia and Dinea became thus coheiresses, and Oppianicus saw almost within his reach the accumulated wealth of the family.

At this moment a stranger appeared in Larino who brought news that the elder brother, Marcus Aurius, was still alive. He had not been killed, as report had said, but had been taken prisoner, and was confined with hard labour at a convict station in the North of Italy. The story was not improbable, and the new-comer produced credible evidence of the truth of what he said. He gave Dinea the names and addresses of persons who had seen Marcus Aurius, and could find him. The hope that she had still a son surviving came to comfort her in her desolation, and she despatched friends to discover him, purchase his release, and restore him to her.

So unpleasant a discovery came inopportunely for the schemes of Oppianicus; but he lost neither heart nor presence of mind. He made acquaintance with the stranger, purchased his help, and induced him to vary his account, and throw Dinea on a false scent. He sent off a confederate to gain the parties in the North and mislead the mother's messengers, while a certain Sextus Vibrius was despatched to obtain true directions from them, to find out Marcus Aurius, and assassinate him. The game was dangerous, however, as long as Dinea lived. She had Aurian kinsmen in Larino who were powerful, and to whom she might possibly appeal. He was aware that her suspicions would turn upon himself as soon as she should hear that her son could not be found, and he thought it better to anticipate future trouble by removing her at once. She was growing old, and her health had been shaken by sorrow and anxiety. Oppianicus recommended to her the assistance of a physician of whose skill he professed to have experience. Dinea declined his advice, and sent for another doctor from Ancona, whom Oppianicus had some difficulty in gaining over to his purpose. He succeeded at last, however, with a bribe of four thousand pounds, and the unfortunate woman was poisoned. Before she died she, too, made a will; but Oppianicus destroyed it. His agents in the north sent him word that his work had been successfully done. Marcus Aurius had been found and killed, and all traces were destroyed by which his fate could be discovered. Oppianicus at once divorced his present wife, married Magia, and took possession of the estates in her name.

He had played his cards skillfully; but again, as with his adventure at Rome, without having succeeded perfectly in averting suspicion from himself. Many eyes, no doubt, were watching him. The Larinates could not see with complaisance the entire disappearance of one of their most honoured families, and the Aurian estates passing into the hands of a blemished and bankrupt adherent of the Oligarchic faction. The messengers sent by Dinea reported that they could not discover Marcus Aurius; but they had found that secret efforts had been made to baffle them. They had ascertained that Oppianicus had been concerned in those efforts, and they wrote to Larino, charging him with foul play. Dinea being dead, the letters were taken to the nearest relative of the family, Aurius Melinus.

This Aurius Melinus had already appeared before the Larinate public in a not very creditable manner. Soon after the death of her father he had married Cluentia, daughter of the widow Sassia, and sister of Aulus Cluentius Avitus. Sassia, who was a licentious, unprincipled woman, became enamoured of her son-in-law. Under the ancient Roman law, the marriage tie had been as indissoluble as in the strictest Christian community. But the restraint of marriage, like every other check on the individual will, had gone down before the

progress of democracy. To divorce a wife was now as easy as
to change a dress. The closest affinity was no longer an
obstacle to a new connection. Sassia succeeded in enchanting
her son-in-law. The daughter was divorced, and the mother
was installed in her place.

Public opinion, though degenerate, was not entirely corrupted. The world of Larino considered itself outraged by
what it still regarded as incest. Aulus Cluentius, the son, took
his mother's conduct so much to heart that he refused to
see either her or her husband, and the domestic scandal had
created almost as much agitation as the tragedy of Dinea and
her children. The two vicious streams were now to unite.
Aurius Melinus, perhaps to recover the esteem of his fellow-
citizens, put himself forward to demand justice against the murderers of his kinsmen. He called a public meeting; he read
aloud in the assembly the letters from the North denouncing
Oppianicus. He demanded an immediate investigation. If
his cousin Marcus was no longer alive, he charged Oppianicus
with having assassinated him.

Suspicions already rife turned to certainty. The people
rose. They rushed to Oppianicus's house to seize and tear him
in pieces. Exceptional villains appear at times to be the special
care of Providence, as if they had a work given them to do
and might not perish till it was accomplished. Oppianicus had
fled; and unhappily a political revolution had not only
provided him with a sure refuge, but with means yet more
fatal of adding to his crimes. While Sylla was fighting
Mithridates in Asia, Marius had returned to a seventh Consul-
ship, and the democracy had enjoyed a brief and sanguinary
triumph; but Marius was dead, and Sylla had returned a
conqueror, and the name of every eminent advocate of popular
rights had now entered on a proscription list. Sylla's lieu-
tenant, Quintus Metellus, was encamped not far from Larino.
Oppianicus threw himself on Metellus's protection, representing himself, perhaps, as the victim of a popular commotion.
Metellus sent him on to the Dictator, and from Sylla he
received a commission to purge Larino of its suspected
citizens, to remove the magistrates, and to execute every one
who had been connected with the Marian faction. In the

haste of the time he was allowed to draw the list of the proscribed himself, and to enter upon it both his open enemies and the accomplices of his crimes, whose too intimate acquaintance with him he had reason to fear. Aurius Melinus perished, and every remaining member of the Aurian kindred. Sextus Vibrius perished, who had been his instrument in hiding the traces of Marcus Aurius and murdering him. The proscribed were seized and killed without being allowed to speak; and thus at one blow Oppianicus was able to rid himself of every one whose vengeance he had to fear, and of the only witness by whom the worst of his crimes could be brought home to him.

For his services to Sylla he was probably rewarded further out of the estates of his victims, and by a series of enormous crimes, which even in that bad time it is to be hoped could not be easily paralleled, he had become the most opulent and most powerful citizen of his native town.

Oppianicus had obtained all that he had desired, but he found, as all mortals find, that the enjoyment had been in the pursuit—that the prize when won still failed to give perfect satisfaction. Happiness was still flying before him— almost within his grasp, but still eluding it. Perhaps the murder of her husband, her mother, and her brothers, may have sate uneasily upon Magia. At any rate he had grown weary of Magia. She too was now cleared away to make room for a more suitable companion. On the death of Aurius Melinus, Sassia was again a widow, and Oppianicus became a suitor for her hand. It was true that he had killed her husband, but he swore, like Richard, that he had done it ' to help her to a better husband.' It was Sassia's ' heavenly face ' which had set him on, and Sassia listened, not unfavour- ably. There were difficulties, however, which had first to be removed. Sassia was rich, and in a position to make conditions. Oppianicus had three children, whose mothers she may have disliked, or whom she expected that she would find in her way. She was willing to tolerate the eldest, who bore his father's name, but she refused to marry him till the two little ones had been removed.

The horrible woman was showing herself a suitable mate

for Oppianicus. Her wealth, her person, perhaps this last proof of the hardiness of her disposition, determined him to secure her on her own terms. One of his little boys was being brought up with his mother at Theano. He sent for the child to Larino. In the night it was taken ill and died, and to prevent inquiry into the manner of its death, the body was burnt before dawn the next morning. Two days after the other little boy died with as mysterious suddenness; and Sassia became Oppianicus's wife.

The people of Larino shuddered and muttered. They could not challenge the favourite of Sylla, the chief magistrate of the town, who had the local authority in his hands and the confidence of the Dictator of Rome; but they shrank from contact with him. They avoided both him and his wife as if they had the plague. Young Cluentius especially held aloof from his mother more sternly than ever, and would neither speak to her nor see her.

At length Sylla died; the middle classes through Italy drew their breath freely again, and at Larino as elsewhere the people could venture to make their voices heard. There was in the town an ancient and venerable college of Priests of Mars, a sort of Cathedral Chapter. The priests had obtained the Roman franchise as a result of the Italian war. It had been confirmed to them by Marius. It had been taken away again by Sylla. And now that Sylla was gone, a deputation from the town was sent to the Senate to petition for its restoration. With this deputation, as one of its members, went young Aulus Cluentius, who was then acquiring fame as a public speaker, and he soon attracted notice at Rome by his vindication of the rights of the Chapter. Oppianicus, who had been Sylla's instrument in carrying out the disfranchisement in Larino, had his own good reasons for dreading to see his work overthrown. With the restoration of political liberty municipal self-govern-ment would be restored along with it. He feared Cluentius on personal grounds as well as political. He saw in him his future accuser, and he had a further motive of another kind for wishing to destroy him. Cluentius had not yet made his will, for he would not leave his fortune to his mother, and he could not bring himself to make a disposition in which her name should not be mentioned. In the absence of a will she was his heir-at-law. It was but one more murder, and Oppianicus would at once quit himself of a dangerous antagonist, gratify his wife, and add the lands of the Cluentii to the vast estates which he had accumulated already.

Cluentius was out of health. Cleophantus, the physician by whom he was attended, was a man of eminence and character, whom it was unsafe to approach by the means which he had used so successfully in the poisoning of Dinea. But Cleophantus had a slave who worked in his laboratory, whom Oppianicus calculated on finding corruptible, and the assistant by whom medicines are made up is in such cases as useful as his principal. He did not think it prudent to appear in person, but a patrician friend, one of the Fabricii, undertook the business for him; and Fabricius felt his way with the slave through his freedman Scamander.

Villains have an instinct for recognising one another, and rarely make mistakes in the character of the persons whom they address. The necessary tact, however, was wanting to Scamander; and in the class of wretches who were bought like sheep in the market, and might be flung at pleasure into the fishponds to feed the aristocrats' lampreys, a degree of virtue was found at last which was to bring Oppianicus's atrocities to a close. Diogenes—so the slave was called—received Scamander's overtures with apparent acquiescence. He listened, drew Scamander on to reveal the name of his employers, and then whispered the story to his master. Cleophantus carried it to Cluentius. An honest senator, Marcus Bibrius, was taken into counsel; and it was agreed that Oppianicus should be played with till he had committed himself, when punishment could at last overtake him. Diogenes kept up his correspondence with Scamander, and promised to administer the poison as soon as he was provided with materials. It was arranged that Cluentius should purchase Diogenes, that he might have a skilled attendant to wait upon him in his illness. The conspiracy would then be carried on under Cluentius's own roof, where the proceedings could be conveniently watched, and the conversations overheard. Oppianicus was out-manœuvred at last. Both he and Fabricius were tempted to betray themselves. The poison was conveyed to Diogenes; the money which was to pay for the murder was brought to him, and received in the presence of concealed witnesses. The criminals were caught red-handed, without room for denial or concealment. They were seized and denounced, and brought to immediate trial.

Horrible crimes have, unfortunately, been so frequent in this world that they have no permanent interest for us; and, unless they have been embalmed in poetry, or are preserved by the exceptional genius of accomplished historians, the memory of them rarely survives a single generation. The tragedies of Larino would have passed into oblivion with the lives of those who had witnessed and shuddered at them. Posterity, if it cared to recollect, would have had their curiosity and their sense of justice satisfied if they could have learned that the chief villain was detected and punished at last; and to revive an interest in a detailed chapter of human wickedness after nearly two thousand years would have been alike super- fluous and impossible. The story, however, now assumes features of deeper importance. Oppianicus and his victims are nothing to us. The rise and fall of the Roman Common- wealth is of undying consequence to the political student; and other thousands of years will still have to pass before we shall cease to study the most minute particulars which will interpret to us so remarkable a phenomenon. The judicial investigation into the crimes of Oppianicus was to form an illustration of the incurable corruption of the Roman Senate; and that Senate's most brilliant member—better known to English schoolboys than the most distinguished modern classic (Kikero they now call him; but we are too old to learn the new nomenclature)—was to be the principal instrument in exposing it.

Criminal trials at Rome were conducted before a body of judges or jurymen, the selection of whom had been one of the chief subjects of contention during the recent political struggles. The privileged orders affected to fear that justice would be degraded if the administration of it was extended to persons who were incompetent for so honourable an office. The people complained that their lives and properties were

unsafe in the hands of proud, extravagant, and venal aristocrats. The Senators declared that if members of their own
order had not been always pure, the middle classes would be
found immeasurably worse. The middle classes, without laying
claims to superior virtue, protested that the Senators had
already descended to the lowest depths of the abyss of
dishonesty.

That the office of a judge, at any rate, might be made one
of the most lucrative situations which the State had to offer
became apparent in a prosecution which happened about the
same time of the Prætor Verres for the plunder of Sicily. In
the trial of Verres it was proved that the governor of a Roman
province under the Republic, looked on his period of office as
an opportunity of making his fortune by extortion and the
public sale of justice. To be successful, he must carry off three
times as much booty as he expected to be allowed to retain.
A third had to be bestowed in buying the goodwill of the
consuls, tribunes, and other magistrates; a third in corrupting
the juries, when he was called to account by the pillaged
provincials; the remaining part only he might calculate on
keeping for himself.

The Court which was to try the case of the Larinates was
composed of thirty-two Senators. Caius Gracchus had granted
the jury-right to the Equites; but it had again been taken
from them by Sylla. The judges were now exclusively
patricians, the purest blood of which Rome had to boast.
Scamander, Fabricius, and Oppianicus were indicted success-
ively for conspiring the murder of Cluentius. The prisoners
were tried separately. Though rumour had caught hold of
some features of the story, the circumstances were generally
unknown. Oppianicus, through his wealth and connections,
had secured powerful patrons; and Cicero, who rarely took
part in prosecutions, was retained in the first instance to
defend Scamander.

Publius Canutius opened the case for Cluentius; and
Cicero, though he exerted himself to the utmost, very soon
discovered that he had a bad cause. The evidence was
absolutely conclusive. Scamander was condemned, and
Fabricius was brought to the bar. Cicero withdrew from the

case and contented himself with watching it. Fabricius's brother, Cepasius, took his place as advocate; but with no better success. Fabricius, too, was convicted, but with a slight difference in the form of the result. A unanimous verdict was given against Scamander; a single Senator, called Stalenus, voted for the acquittal of Fabricius. There was no more doubt of his guilt than of his freedman's. The evidence against them both was the same. Stalenus had not been bribed, for Fabricius was poor; but he intended to intimate to the rich Oppianicus that he was open to an arrangement when his own turn should come on.

Stalenus was a man of consequence. He had been quæstor, and aspired to the higher offices of State. He had obtained some notoriety in a recent civil case in which one of the parties was a certain Safinius Atella. Safinius had the worst of the argument, and Stalenus had boasted that for a round sum of money he could purchase a verdict notwithstanding. The money was given to him, but Safinius lost his cause, and ill-natured persons had whispered that Stalenus had kept it for himself. Such a transaction, however, if successful and unde-tected, might pass for a stroke of cleverness. At all events the suspicions attached to it had not interfered with the further employment of this ingenious young nobleman. He was merely observed, and anything singular in his conduct was set down to its right motive.

Oppianicus's case might well be considered desperate. Scamander and Fabricius had been accessories only to a single attempt at murder. The past history of Oppianicus had probably been alluded to generally in the preliminary trials. He would stand at the bar an object of general abhorrence for various other enormities, and the proofs which had been sufficient to condemn his accomplices would tell with tenfold force against their instigator, whose past career had been so dark. In the vote of Stalenus only some glimmer of hope remained. The Court adjourned for a few days. In the interval Oppianicus made Stalenus's acquaintance, and they soon understood one another. Stalenus told him frankly that his situation was a difficult one, and would probably be expensive. The judges who had condemned the other prisoners would commit manifest perjury if they acquitted Oppianicus. Public feeling being excited, they would be exposed to general opprobrium, and they would require to be well paid for their services. Still, however, he thought it might be managed. He knew his men, and he considered that he could secure fifteen votes out of the thirty-two, which in addition to his own would be sufficient. Money only was necessary : each vote would require £400.

Oppianicus's fortune would be of little use to him if he was convicted. Being a Roman citizen, he was not liable to a sentence of death from a criminal court, but exile and a fine amounting nearly to confiscation were as bad or possibly worse. He assented to Stalenus's terms, and paid into his hands £6400.

It was understood by this time that a negotiation with the prisoner was going forward. Stalenus had felt his way, dropping hints here and there in whatever quarter they were likely to be operative, and at length the corruptible fifteen had given conditional assurances that they might be relied on. But the terms, as he expected, were high; very little would be left for himself; and he began to reflect that with perfect safety he might keep the whole of it. The honest part of the jury would, he thought, undoubtedly vote for a conviction. Those who had agreed to sell their consciences would be so angry if they were now disappointed that he might count on them with equal certainty, and it would be in vain that after a verdict of guilty such a wretch as Oppianicus would appeal to public opinion. No one would believe him, no one would pity him. Thus the night before the trial came on he informed his friends upon the jury that Oppianicus had changed his mind, and that no money was forthcoming. They were as exasperated as he hoped to find them. He was himself not suspected, and they met the next day in court with a most virtuous resolution that justice should not be baulked of its object.

The voting in a Roman trial was either open or secret, as the Court might decide for itself. Oppianicus not relying too perfectly on his friends, and anxious not to be cheated of the wares for which he had paid, demanded that each judge should give his verdict by word of mouth. The tribune Quinctius, who was secretly his friend, supported him, and his request was agreed to. Every one was aware that there had been bribery, and the members of the jury who were open to bribes were generally well known. It was, of course, assumed that they would vote for an acquittal, and Stalenus and his friends were observed with contemptuous curiosity, but without a doubt of what their judgment would be.

It happened that Stalenus was the first to vote, and two of his intimate associates were the second and third. To the astonishment of every one, all three without the slightest hesitation voted guilty. The rest of the judges, or rather the respectable portion of them, were utterly bewildered. The theory of corruption implies that men who take bribes will generally fulfil their contract, nor again do men usually take bribes to vote according to their real convictions. They were assured that Stalenus had been corrupted by some one to give a false verdict. They thought he had been corrupted by Oppianicus; but he had voted against Oppianicus; he had voted for Cluentius,—therefore it seemed he must have been bribed by Cluentius, and Oppianicus might be innocent after all. Thus argued the outside public almost universally, having heard the story but imperfectly. Thus argued even a section of the judges themselves, and in their confusion five of the more honest of them actually voted for Oppianicus's acquittal. The larger number concluded at last that they must go by the evidence. Stalenus and his friends might have taken money from Cluentius. Cluentius might have been afraid to trust himself entirely to the justice of his cause. But corruption could not alter the truth. Oppianicus was unquestionably guilty, and he was condemned by a large majority.

He for his part was banished, clamouring that he was betrayed, but unable, as Stalenus expected, to obtain a remission of his sentence. In modern eyes such a punishment was immeasurably too lenient. To a Roman who wanted courage to end his misfortune with his own hand, exile was held to be the most terrible of calamities. Cæsar pleaded against the execution of the accomplices of Catiline, that death ended all things. He would have them live and suffer. ' Life,' said

Cicero on the present occasion, 'was worse than death to Oppianicus. No one believed any longer the old wives' fable of Tartarus. Death would be but a happy release to him.' He left Rome to wander about Italy, as if marked with a curse. Sassia followed him to torment him with her reproaches and infidelities. One day as he was riding his horse threw him. He was mortally injured and died.

So ended Oppianicus. So, however, did not end the consequences of his various villanies. Political passions were again rising. The people in Rome and out of it were clamouring to the skies against the iniquities of the Senate. The story went abroad that a senatorial jury had again been bribed; and being without detailed knowledge of the case, the Roman populace rushed naturally to the conclusion that an innocent man had been condemned. Oppianicus had protested against the verdict, and had denounced his judges. It was enough. The verdict was indisputably corrupt, and a corrupt verdict, as a matter of course, must be a false verdict.

Quinctius the tribune, Oppianicus's friend, encouraged the agitation. It was an opportunity not to be neglected of bringing the Senate into disrepute. Thrice he harangued the General Assembly in the Forum. He insisted that the degraded patricians should be stripped once more of the privileges which they abused. Cluentius's name became a by-word. He who in his humble way had been the champion of his own townspeople was identified with the hated senatorial monopoly. So furious were the people that for eight years, Cicero says, they would not so much as listen to a word that could be said for him. Every senator who had voted for Oppianicus's condemnation was prosecuted under the Jury Laws. Some were fined, some were expelled from the Senate by the Censors. One of them, Caius Egnatius, was disinherited by his father. The Senate itself was invited to condemn its own members. Not daring to refuse, the Senate saved its conscience by a wise generality, and passed a resolution that any person or persons who had been instrumental in corrupting public justice had been guilty of a heinous offence. Finally Cluentius himself was brought to trial, and so hot was public feeling against him that Cicero was obliged to confine his defence to a legal technicality. The law, he said, was for the restraint of corruption of the juries. The juries under Sylla's constitution could consist of senators only, and Cluentius being an Eques, the law could not touch him.

Gradually the outcry died away, melting into the general stream of indignation which in a few years swept away the constitution, and under new forms made justice possible again. But the final act of the Cluentian drama had still to be played out. Again Cluentius was to appear before a tribunal of Roman judges. Again Cicero was to defend him—no longer under a quibble, but on the merits of the whole case, into which at least it was possible to enter.

From the speech which Cicero delivered on this occasion we have gathered our story. It is not a favourable specimen of his oratorical power. There is no connection in the events. There is no order of time. We are hurried from date to date, from place to place. The same person is described under different names; the same incident in different words. The result is a mass of threads so knotted, twisted, and entangled, that only patient labour can sort them out into intelligible arrangement.

What Cicero lacks in method, however, he makes up in earnestness. He was evidently supremely affected by the combination of atrocities and misunderstandings by which an innocent, well-deserving man was likely to be overwhelmed.

The various lovers of Sassia had been either murdered or had died, or had deserted her. She had lost much of her ill-gained fortunes. She had grown too old for the further indulgence of her pleasant vices. One desire alone remained, and had devoured the rest—a desire for revenge upon her son Cluentius. In the prejudiced condition of public feeling at Rome, any wily accusation against him might be expected to obtain a hearing. Having escaped the prosecution for the bribery of the judges, he was charged with having murdered one of his friends, whose property he hoped to inherit. The attempt was clumsy and it failed. The friend was proved to have died where Cluentius could have had no access to him; and a nephew, and not Cluentius, was his heir. The next accusation was of having tried to poison the surviving son of Oppianicus. Cluentius and the younger Oppianicus had been together at a festival of Larino. Another youth who was also present there had died a few days later, and it was alleged that he had drunk by mistake from a cup which had been prepared for Sassia's stepson. But again the evidence broke down. There was no proof that the death was caused by poison, or that Cluentius was in any way connected with it.

The accursed woman, though twice baffled, would not abandon her object. In both instances proof of malice had been wanting. Cluentius had no object in perpetuating either of the crimes of which she had accused him. If he had no grudge against the young Oppianicus, however, he had undoubtedly hated his father, and she professed to have discovered that the father had not died, as had been reported, by the fall from his horse, but had been poisoned by a cake which had been administered to him at Cluentius's instigation. The method in which Sassia went to work to make out her case throws a fresh and hideous light on the Roman administration of justice in the last days of liberty. She produced two witnesses who were both slaves. To one of them, Nicostratus, a Greek, she owed an old grudge. He had belonged to Oppianicus the elder, and had revealed certain infidelities of hers which had led to inconvenience. The other, Strato, was the slave of a doctor who had attended Oppianicus after his accident. Since neither of these men were willing to say what she required them to say of their own accord, she demanded according to custom that they should be tortured. The Roman law did not acknowledge any rights in these human chattels; a slave on the day of his bondage ceased to be a man. Nicostratus and Strato were racked till the executioners were weary, but nothing could be extracted from them. A distinguished advocate who was present, and was not insensible to pity, said that the slaves were being tortured not to make them tell the truth, but to make them lie. The court took the same view, and they were released.

Once more Sassia was defeated, but she waited her opportunity. Three years later, the orator Hortensius, a general protector of rogues, was elected to the consulate. The vindictiveness with which she had come forward as the prosecutrix of her own son had injured her cause. She made one more effort, and this time she prevailed on the young Oppianicus, who had meanwhile married her daughter, to appear in her place. She had purchased Strato after his escape from the torture, and had power of life and death over him. He had murdered a fellow slave; and it was alleged that when he confessed to this crime he had confessed to the other also. He was crucified, and to prevent his telling inconvenient truths upon the cross, his tongue was cut out before he was nailed upon it. On the strength of his pretended deposition, a criminal process was once more instituted against Cluentius before a Roman jury. The story had by this time become so notorious, and the indignation of the provinces had been so deeply roused, that deputations from every town in the south of Italy came to the Capital to petition in Cluentius's favour. How the trial ended is unknown. It may be hoped that he was acquitted—but it is uncertain. Innocent men have suffered by millions in this world. As many guilty wretches have escaped, and seemed to triumph; but the vengeance which follows upon evil acts does not sleep because individuals are wronged. The penalty is exacted to the last farthing from the community which permits injustice to be done. And the Republican Commonwealth of Rome was fast filling the measure of its iniquities. In another half-century perjured juries and corrupted magistrates had finished their work; the world could endure them no longer, and the free institutions which had been the admiration of mankind were buried under the throne of the Cæsars.

A BISHOP OF THE TWELFTH
CENTURY[1]

To the sceptical student of the nineteenth century the
ecclesiastical biographies of mediæval Europe are for the most
part unprofitable studies. The writers of them were generally
monks. The object for which they were composed was either
the edification of the brethren of the convent, or the glorifying
of its founder or benefactor. The Holy See in considering a
claim to canonization disregarded the ordinary details of
character and conduct. It dwelt exclusively on the exceptional
and the wonderful, and the noblest of lives possessed but little
interest for it unless accompanied by evidences of miracles,
performed directly by the candidate while on earth or by his
relics after his departure. Instead of pictures of real men
the biographers present us with glorified images of what,
in their opinion, the Church heroes ought to have been.
St Cuthbert becomes as legendary as Theseus, and the authentic
figure is swathed in an embroidered envelope of legends
through which usually no trace of the genuine lineaments is
allowed to penetrate.

It happens however, occasionally, that in the midst of the
imaginative rubbish which has thus come down to us, we
encounter something of a character entirely different. We find
ourselves in the hands of writers who themselves saw what
they describe, who knew as well as we know the distinction
between truth and falsehood, and who could notice and
appreciate genuine human qualities. Amidst the obscure
forms of mediæval history we are brought face to face with
authentic flesh and blood, and we are able to see in clear
sunlight the sort of person who, in those ages, was considered
especially admirable, and, alive or dead, was held up to the
reverence of mankind. To one of these I propose in the

[1] *Magna Vita S. Hugonis Episcopi Lincolnensis.* From MSS. in
the Bodleian Library, Oxford, and the Imperial Library, Paris. Edited
by the Rev. James F. Dimock, M.A., Rector of Bamburgh, Yorkshire.

present article to draw some brief attention. It is the life
of St Hugo of Avalon, a monk of the Grand Chartreuse[2], who
was invited by Henry II into England, became Bishop of
Lincoln, and was the designer, and in part builder, of Lincoln
Cathedral. The biographer was his chaplain and constant
companion—Brother Adam—a monk like himself, though of
another order, who became afterwards Abbot of Ensham[3]; and
having learnt, perhaps from the Bishop himself, the detest-
ableness of lying, has executed his task with simple and
scrupulous fidelity. The readers whose interests he was
considering were, as usual, the inmates of convents. He
omits, as he himself tells us, many of the outer and more
secular incidents of the Bishop's life, as unsuited to his
audience. We have glimpses of kings, courts, and great
councils, with other high matters of national moment. The
years which the Bishop spent in England were rich in events.
There was the conquest of Ireland; there were Welsh and
French wars; the long struggle of Henry II and his sons; and,
when Henry passed away, there was the Grand Crusade. Then
followed the captivity of Cœur de Lion and the treachery of
John; and Hugo's work, it is easy to see, was not confined
to the management of his diocese. On all this, however, Abbot
Adam observes entire silence, not considering our curiosity, but
the concerns of the souls of his own monks, whom he would
not distract by too lively representations of the world which
they had abandoned.

The book however, as it stands, is so rare a treasure that
we will waste no time in describing what it is not. Within its
own compass it contains the most vivid picture which has come
down to us of England as it then was, and of the first
Plantagenet kings.

Bishop Hugo came into the world in the mountainous
country near Grenoble, on the borders of Savoy. Abbot Adam
dwells with a certain pride upon his patron's parentage. He
tells us indeed, sententiously, that it is better to be noble
in morals than to be noble in blood—that to be born
undistinguished is a less misfortune than to live so—but he
regards a noble family only as an honourable setting for a

[2] [Grande Chartreuse.] [3] [Eynsham, in Oxfordshire.]

nature which was noble in itself. The Bishop was one of three children of a Lord of Avalon, and was born in a castle near Pontcharra. His mother died when he was eight years old; and his father having lost the chief interest which bound him to life, divided his estates between his two other sons, and withdrew with the little one into an adjoining monastery. There was a college attached to it, where the children of many of the neighbouring barons were educated. Hugo, however, was from the first designed for a religious life, and mixed little with the other boys. 'You, my little fellow,' his tutor said to him, "I am bringing up for Christ: you must not learn to play or trifle.' The old Lord became a monk. Hugo grew up beside him in the convent, waiting on him as he became infirm, and smoothing the downward road; and meanwhile learning whatever of knowledge and practical piety his preceptors were able to provide. The life, it is likely, was not wanting in austerity, but the comparatively easy rule did not satisfy Hugo's aspirations. The theory of 'religion,' as the conventual system in all its forms was termed, was the conquest of self, the reduction of the entire nature to the control of the better part of it; and as the seat of self lay in the body, as temptation to do wrong, then as always, lay directly or indirectly, in the desire for some bodily indulgence, or the dread of some bodily pain, the method pursued was the inuring of the body to the hardest fare, and the producing indifference to cold, hunger, pain, or any other calamity which the chances of life could inflict upon it. Men so trained could play their part in life, whether high or low, with wonderful advantage. Wealth had no attraction for them. The world could give them nothing which they had learnt to desire, and take nothing from them which they cared to lose. The orders, however, differed in severity; and at this time the highest discipline, moral and bodily, was to be found only among Carthusians. An incidental visit with the prior of his own convent to the Grande Chartreuse, determined Hugo to seek admission into this extraordinary society.

It was no light thing which he was undertaking. The majestic situation of the Grande Chartreuse itself, the loneliness, the seclusion, the atmosphere of sanctity, which hung around it, the mysterious beings who had made their home there, fascinated his imagination. A stern old monk, to whom he first communicated his intention, supposing that he was led away by a passing fancy, looked grimly at his pale face and delicate limbs, and roughly told him that he was a fool. ' Young man,' the monk said to him, ' the men who inhabit these rocks are hard as the rocks themselves. They have no mercy on their own bodies and none on others. The dress will scrape the flesh from your bones. The discipline will tear the bones themselves out of such frail limbs as yours.'

The Carthusians combined in themselves the severities of the hermits and of the regular orders. Each member of the fraternity lived in his solitary cell in the rock, meeting his companions only in the chapel, or for instruction, or for the business of the house. They ate no meat. A loaf of bread was given to every brother on Sunday morning at the refectory door, which was to last him through the week. An occasional mess of gruel was all that was allowed in addition. His bedding was a horse-cloth, a pillow, and a skin. His dress was a horsehair shirt, covered *outside* with linen, which was worn night and day, and the white cloak of the order, generally a sheepskin, and unlined—all else was bare. He was bound by vows of the strictest obedience. The order had business in all parts of the world. Now some captive was to be rescued from the Moors; now some earl or king had been treading on the Church's privileges; a brother was chosen to interpose in the name of the Chartreuse : he received his credentials and had to depart on the instant, with no furniture but his stick, to walk, it might be, to the furthest corner of Europe.

A singular instance of the kind occurs incidentally in the present narrative. A certain brother Einard, who came ultimately to England, had been sent to Spain, to Granada, to Africa itself. Returning through Provence he fell in with some of the Albigenses, who spoke slightingly of the sacraments. The hard Carthusian saw but one course to follow with men he deemed rebels of his Lord. He was the first to urge the crusade which ended in their destruction. He roused the nearest orthodox nobles to arms, and Hugo's biographer tells delightedly how the first invasions were followed up by others on a larger scale, and 'the brute and pestilent race, unworthy of the name of men, were cut away by the toil of the faithful, and by God's mercy destroyed.'

'Pitiless to themselves,' as the old monk said, 'they had no pity on any other man,' as Einard afterwards was himself to feel. Even Hugo at times disapproved of their extreme severity. 'God,' he said, alluding to some cruel action of the society, 'God tempers his anger with compassion. When he drove Adam from Paradise, he at least gave him a coat of skins : man knows not what mercy means.'

Einard, after this Albigensian affair, was ordered in the midst of a bitter winter to repair to Denmark. He was a very aged man—a hundred years old, his brother monks believed—broken at any rate with age and toil. He shrank from the journey, he begged to be spared, and when the command was persisted in, he refused obedience. He was instantly expelled. Half-clad, amidst the ice and snow, he wandered from one religious house to another. In all he was refused admission. At last, one bitter frosty night he appeared penitent at the gate of the Chartreuse, and prayed to be forgiven. The porter was forbidden to open to him till morning, but left the old man to shiver in the snow through the darkness.

'By my troth, brother,' Einard said the next day to him, 'had you been a bean last night, between my teeth, they would have chopped you in pieces in spite of me.'

Such were the monks of the Chartreuse, among whom the son of the Avalon noble desired to be enrolled, as the highest favour which could be shown him upon earth. His petition was entertained. He was allowed to enlist in the spiritual army, in which he rapidly distinguished himself; and at the end of twenty years he had acquired a name through France as the ablest member of the world-famed fraternity.

It was at this time, somewhere about 1174, that Henry II conceived the notion of introducing the Carthusians into England. In the premature struggle to which he had committed himself with the Church, he had been hopelessly worsted. The Constitutions of Clarendon had been torn in pieces. He had himself, of his own accord, done penance at the shrine of the murdered Becket. The haughty sovereign of England, as a symbol of the sincerity of his submission, had knelt in the Chapter-house of Canterbury, presenting voluntarily there his bare shoulders to be flogged by the monks. His humiliation, so far from degrading him, had restored him to the affection of his subjects, and his endeavour thenceforward was to purify and reinvigorate the proud institution against which he had too rashly matched his strength.

In pursuance of his policy he had applied to the Chartreuse for assistance, and half a dozen monks, among them brother Einard, whose Denmark mission was exchanged for the English, had been sent over and established at Witham, a village not far from Frome in Somersetshire. Sufficient pains had not been taken to prepare for their reception. The Carthusians were a solitary order and required exclusive possession of the estates set apart for their use. The Saxon population were still in occupation of their holdings, and being Crown tenants, saw themselves threatened with eviction in favour of foreigners. Quarrels had arisen and ill-feeling, and the Carthusians, proud as the proudest of nobles, and considering that in coming to England they were rather conferring favours than receiving them, resented the being compelled to struggle for tenements which they had not sought or desired. The first prior threw up his office and returned to the Chartreuse. The second died immediately after of chagrin and disgust; and the King, who was then in Normandy, heard to his extreme mortification that the remaining brethren were threatening to take staff in hand and march back to their homes. The Count de Maurienne to whom he communicated his distress mentioned Hugo's name to him. It was determined to send for Hugo, and Fitzjocelyn, Bishop of Bath, with other venerable persons carried the invitation to the Chartreuse.

To Hugo himself, meanwhile, as if in preparation for the destiny which was before him, a singular experience was at that moment occurring. He was now about forty years old. It is needless to say that he had duly practised the usual austerities prescribed by his rule. Whatever discipline could do to kill the carnal nature in him had been carried out to its utmost harshness. He was a man, however, of great physical strength. His flesh was not entirely dead, and he was going where superiority to worldly temptation would be specially required. Just before Fitzjocelyn arrived he was assailed suddenly by emotions so extremely violent that he said he would rather face the pains of Gehenna than encounter them again. His mind was unaffected, but the devil had him at advantage in his sleep. He prayed, he flogged himself, he fasted, he confessed; still Satan was allowed to buffet him, and though he had no fear for his soul, he thought his body would die in the struggle. One night in particular the agony reached its crisis. He lay tossing on his uneasy pallet, the angel of darkness trying with all his allurements to tempt his conscience into acquiescence in evil. An angel from above appeared to enter the cell as a spectator of the conflict. Hugo imagined that he sprang to him, clutched him, and wrestled like Jacob with him to extort a blessing but could not succeed, and at last he sank exhausted on the ground. In the sleep or the unconsciousness which followed, an aged prior of the Chartreuse who had admitted him as a boy to the order, had died and had since been canonized, seemed to lean over him as he lay and inquired the cause of his distress. He said that he was afflicted to agony by the law of sin that was in his members, and unless some one aided him he would perish. The saint drew from his breast what appeared to be a knife, opened his body, drew a fiery mass of something from the bowels, and flung it out of the door. He awoke and found that it was morning and that he was perfectly cured.

'Did you ever feel a return of these motions of the flesh?' asked Adam, when Hugo related the story to him.

'Not never,' Hugo answered, 'but never to a degree that gave me the slightest trouble.'

'I have been particular,' wrote Adam afterwards, 'to relate this exactly as it happened, a false account of it having gone abroad that it was the Blessed Virgin who appeared instead of the prior,' and that Hugo was relieved by an operation of a less honourable kind.

Visionary nonsense the impatient reader may say; and had Hugo become a dreamer of the cloister, a persecutor like St Dominic, or a hysterical fanatic like Ignatius Loyola, we might pass by it as a morbid illusion. But there never lived a man to whom the word morbid could be applied with less propriety. In the Hugo of Avalon with whom we are now to become acquainted, we shall see nothing but the sunniest cheerfulness, strong masculine sense, inflexible purpose, uprightness in word and deed; with an ever-flowing stream of genial and buoyant humour.

In the story of the temptation, therefore, we do but see the final conquest of the selfish nature in him, which left his nobler qualities free to act, wherever he might find himself.

Fitzjocelyn anticipating difficulty had brought with him the Bishop of Grenoble to support his petition. He was received at first with universal clamour. Hugo was the brightest jewel of the order; Hugo could not be parted with for any prince on earth. He himself, entirely happy where he was, anticipated nothing but trouble, but left his superiors to decide for him. At length sense of duty prevailed. The brethren felt that he was a shining light, of which the world must not be deprived. The Bishop of Grenoble reminded them that Christ had left heaven and come to earth for sinners' souls, and that his example ought to be imitated. It was arranged that Hugo was to go, and a few weeks later he was at Witham.

He was welcomed there as an angel from heaven. He found everything in confusion, the few monks living in wattled huts in the forest, the village still in possession of its old occupants, and bad blood and discontent on all hands. The first difficulty was to enter upon the lands without wrong to the people, and the history of a large eviction in the twelfth century will not be without its instructiveness even at the present day. One thing Hugo was at once decided upon, that the foundation would not flourish if it was built upon injustice. He repaired to Henry, and as a first step induced him to offer the tenants (Crown serfs or villeins) either entire enfranchisement or farms of equal value, or any other of the royal manors, to be selected by themselves. Some chose one, some the other. The next thing was compensation for improvements, houses, farm-buildings, and fences erected by the people at their own expense. The Crown, if it resumed possession, must pay for these or wrong would be done. 'Unless your Majesty

satisfy these poor men to the last obol,' said Hugo to Henry,
' we cannot take possession.'

The King consented, and the people, when the Prior carried
back the news of the arrangement, were satisfied to go.

But this was not all. Many of them were removing no
great distance, and could carry with them the materials of their
houses. Hugo resolved that they should keep these things,
and again marched off to the court.

' My Lord,' said Hugo, ' I am but a new comer in your
realm, and I have already enriched your Majesty with a
quantity of cottages and farm-steadings.'

' Riches I could well have spared,' said Henry, laughing.
' You have almost made a beggar of me. What am I to do
with old huts and rotten timber?'

' Perhaps your Majesty will give them to me,' said Hugo.
' It is but a trifle,' he added, when the King hesitated. ' It is
my first request, and only a small one.'

' This is a terrible fellow that we have brought among us,'
laughed the King; ' if he is so powerful with his persuasions,
what will he do if he tries force? Let it be as he says. We
must not drive him to extremities.'

Thus, with the good will of all parties, and no wrong done
to any man, the first obstacles were overcome. The villagers
went away happy. The monks entered upon their lands amidst
prayers and blessings, the King himself being as pleased as any
one at his first experience of the character of Prior Hugo.

Henry had soon occasion to see more of him. He had
promised to build the monks a house and chapel, but between
Ireland, and Wales, and Scotland, and his dominions in
France, and his three mutinous sons, he had many troubles
on his hands. Time passed and the building was not begun,
and Hugo's flock grew mutinous once more; twice he sent
Henry a reminder, twice came back fair words and nothing
more. The brethren began to hint that the Prior was afraid of
the powers of this world, and dared not speak plainly; and one
of them, Brother Gerard, an old monk with high blood in his
veins, declared that he would himself go and tell Henry some
unpleasant truths. Hugo had discovered in his interviews
with him that the King was no ordinary man, ' vir sagacis

ingenii, et inscrutabilis fere animi.' He made no opposition, but he proposed to go himself along with this passionate gentleman and he, Gerard and the aged Einard, who was mentioned above, went together as a deputation.

The king received them as ' cœlestes angelos,'—angels from heaven. He professed the deepest reverence for their characters, and the greatest anxiety to please them, but he said nothing precise and determined, and the fiery Gerard burst out as he intended. Carthusian monks, it seems, considered themselves entitled to speak to kings on entirely equal terms. 'Finish your work or leave it, my Lord King,' the proud Burgundian said. ' It shall no more be any concern to me. You have a pleasant realm here in England, but for myself I prefer to take my leave of you and go back to my desert Chartreuse. You give us bread, and you think you are doing a great thing for us. We do not need your bread. It is better for us to return to our Alps. You count money lost which you spend on your soul's health; keep it then, since you love it so dearly. Or rather, you cannot keep it; for you must die and let it go to others who will not thank you.'

Hugo tried to check the stream of words, but Gerard and Einard were both older than he, and refused to be restrained.

' Regem videres philosophantem :' the King was apparently meditating. His face did not alter, nor did he speak a word till the Carthusian had done.

' And what do you think, my good fellow,' he said at last, after a pause, looking up and turning to Hugo : ' will you forsake me too?'

' My Lord,' said Hugo, ' I am less desperate than my brothers. You have much work upon your hands, and I can feel for you. When God shall please you will have leisure to attend to us.'

' By my soul,' Henry answered, ' you are one that I will never part with while I live.'

He sent workmen at once to Witham. Cells and chapel were duly built. The trouble finally passed away, and the Carthusian priory taking root became the English nursery of the order, which rapidly spread.

Hugo himself continued there for eleven years, leaving it

from time to time on business of the Church, or summoned, as happened more and more frequently, to Henry's presence. The King, who had seen his value, who knew that he could depend upon him to speak the truth, consulted him on the most serious affairs of state, and beginning with respect, became familiarly and ardently attached to him. Witham however remained his home, and he returned to it always as to a retreat of perfect enjoyment. His cell and his dole of weekly bread gave him as entire satisfaction as the most luxuriously furnished villa could afford to one of ourselves; and long after, when he was called elsewhere, and the cares of the great world fell more heavily upon him, he looked to an annual month at Witham for rest of mind and body, and on coming there he would pitch away his grand dress and jump into his sheepskin as we moderns put on our shooting jackets.

While he remained Prior he lived in perfect simplicity and unbroken health of mind and body. The fame of his order spread fast, and with its light the inseparable shadow of superstition. Witham became a place of pilgrimage; miracles were said to be worked by involuntary effluences from its occupants. Then and always Hugo thought little of miracles, turned his back on them for the most part, and discouraged them if not as illusions yet as matters of no consequence. St Paul thought one intelligible sentence containing truth in it was better than a hundred in an unknown tongue. The Prior of Witham considered that the only miracle worth speaking of was holiness of life. ' Little I,' writes Adam (parvulus ego), ' observed that he worked many miracles himself, but he paid no attention to them.' Thus he lived for eleven years with as much rational happiness as, in his opinion, human nature was capable of experiencing. When he lay down upon his horse-rug he slept like a child, undisturbed, save that at intervals, as if he was praying, he muttered a composed Amen. When he awoke he rose and went about his ordinary business : cleaning up dirt, washing dishes and such like, being his favourite early occupation.

The Powers, however—who, according to the Greeks, are jealous of human felicity—thought proper, in due time, to disturb the Prior of Witham. Towards the end of 1183 Walter de Coutances was promoted from the Bishopric of Lincoln to the Archbishopric of Rouen. The see lay vacant for two years and a half, and a successor had now to be provided. A great council was sitting at Ensham[4] on business of the realm; the King riding over every morning from Woodstock. A deputation of canons from Lincoln came to learn his pleasure for the filling up of the vacancy. The canons were directed to make a choice for themselves and were unable to agree, for the not unnatural reason that each canon considered the fittest person to be himself. Some one (Adam does not mention the name) suggested, as a way out of the difficulty, the election of Hugo of Witham. The canons being rich, well to do, and of the modern easy-going sort, laughed at the suggestion of the poor Carthusian. They found to their surprise, however, that the King was emphatically of the same opinion, and that Hugo and nobody else was the person he intended for them.

The King's pleasure was theirs. They gave their votes, and despatched a deputation over the downs to command the Prior's instant presence at Ensham.

A difficulty rose where it was least expected. Not only was the ' Nolo episcopari ' in Hugo's case a genuine feeling, not only did he regard worldly promotion as a thing not in the least attractive to him; but, in spite of his regard for Henry, he did not believe that the King was a proper person to nominate the prelates of the Church. He told the canons that the election was void. They must return to their own cathedral, call the chapter together, invoke the Holy Spirit, put the King of England out of their minds, and consider rather the King of kings; and so, and not otherwise, proceed with their choice.

The canons, wide-eyed with so unexpected a reception, retired with their answer. Whether they complied with the spirit of Hugo's direction may perhaps be doubted. They, however, assembled at Lincoln with the proper forms, and repeated the election with the external conditions which he had prescribed. As a last hope of escape he appealed to the Chartreuse, declaring himself unable to accept any office without orders from his superiors; but the authorities there

4 [Eynsham.]

forbade him to decline; and a fresh deputation of canons having come for his escort, he mounted his mule with a heavy heart and set out in their company for Winchester, where the King was then residing.

A glimpse of the party we are able to catch upon their journey. Though it was seven hundred years since, the English September was probably much like what it is at present, and the down country cannot have materially altered. The canons had their palfreys richly caparisoned with gilt saddle-cloths, and servants and sumpter horses. The Bishop elect strapped his wardrobe, his blanket and sheep-skin, at the back of his saddle. He rode in this way resisting remonstrance till close to Winchester, when the canons, afraid of the ridicule of the Court, slit the leathers without his knowing it, and passed his baggage to the servants.

Consecration and installation duly followed, and it was supposed that Hugo, a humble monk, owing his promotion to the King, would be becomingly grateful, that he would become just a Bishop, like anybody else, complying with established customs, moving in the regular route, and keeping the waters smooth.

All parties were disagreeably, or rather, as it turned out ultimately, agreeably, surprised. The first intimation which he gave that he had a will of his own followed instantly upon his admission. Corruption or quasi-corruption had .gathered already round ecclesiastical appointments. The Archdeacon of Canterbury put in a claim for consecration fees, things in themselves without meaning or justice, but implying that a bishopric was a prize, the lucky winner of which was expected to be generous.

The new prelate held no such estimate of the nature of his appointment—he said he would give as much for his cathedral as he had given for his mitre, and left the Archdeacon to his reflections.

No sooner was he established and had looked about him, than from the poor tenants of estates of the see he heard complaints of that most ancient of English grievances—the game laws. Hugo, who himself touched no meat, was not likely to have cared for the chase. He was informed that venison must be provided for his installation feast. He told his people to take from his park what was necessary—three hundred stags if they pleased, so little he cared for preserving them; but neither was he a man to have interfered needlessly with the recognized amusements of other people. There must have been a case of real oppression, or he would not have meddled with such things. The offender was no less a person than the head forester of the King himself. Hugo, failing to bring him to reason with mild methods, excommunicated him, and left him to carry his complaints to Henry. It happened that a rich stall was at the moment vacant at Lincoln. The King wanted it for one of his courtiers, and gave the Bishop an opportunity of redeeming his first offence by asking for it as a favour to himself. Henry was at Woodstock; the Bishop, at the moment, was at Dorchester, a place in his diocese thirteen miles off. On receiving Henry's letter the Bishop made the messenger carry back for answer that prebendal stalls were not for courtiers but for priests. The King must find other means of rewarding temporal services. Henry, with some experience of the pride of ecclesiastics, was unprepared for so abrupt a message—Becket himself had been less insolent—and as he had been personally kind to Hugo, he was hurt as well as offended. He sent again to desire him to come to Woodstock, and prepared, when he arrived, to show him that he was seriously displeased. Then followed one of the most singular scenes in English history—a thing veritably true, which oaks still standing in Woodstock Park may have witnessed. As soon as word was brought that the Bishop was at the park gate, Henry mounted his horse, rode with his retinue into a glade in the forest, where he alighted, sat down upon the ground with his people, and in this position prepared to receive the criminal. The Bishop approached—no one rose or spoke. He saluted the King; there was no answer. Pausing for a moment, he approached, pushed aside gently an earl who was sitting at Henry's side, and himself took his place. Silence still continued. At last Henry, looking up, called for a needle and thread; he had hurt a finger of his left hand. It was wrapped with a strip of linen rag, the end was loose, and he began to sew. The Bishop watched him through a few stitches, and then, with the utmost composure, said to him—'Quam similis es modo cognatis tuis de Falesiâ'—'your Highness now reminds me of your cousins of Falaise.' The words sounded innocent enough—indeed, entirely unmeaning. Alone of the party, Henry understood the allusion; and, overwhelmed by the astonishing impertinence, he clenched his hands, struggled hard to contain himself, and then rolled on the ground in convulsions of laughter.

'Did you hear,' he said to his people when at last he found words; 'did you hear how this wretch insulted us? The blood of my ancestor the Conqueror, as you know, was none of the purest. His mother was of Falaise, which is famous for its leather work, and when this mocking gentleman saw me stitching my finger, he said I was showing my parentage.'

'My good sir,' he continued, turning to Hugo, 'what do you mean by excommunicating my head forester, and when I make a small request of you, why is it that you not only do not come to see me, but do not send me so much as a civil answer?'

'I know myself,' answered Hugo gravely, 'to be indebted to your Highness for my late promotion. I considered that your Highness's soul would be in danger if I was found wanting in the discharge of my duties; and therefore it was that I used the censures of the Church when I held them necessary, and that I resisted an improper attempt on your part upon a stall in my cathedral. To wait on you on such a subject I thought superfluous, since your Highness approves, as a matter of course, of whatever is rightly ordered in your realm.'

What could be done with such a Bishop? No one knew better than Henry the truth of what Hugo was saying, or the worth of such a man to himself. He bade Hugo proceed with the forester as he pleased. Hugo had him publicly whipped, then absolved him, and gave him his blessing, and found in him ever after a fast and faithful friend. The courtiers asked for no more stalls, and all was well.

In Church matters in his own diocese he equally took his own way. Nothing could be more unlike than Hugo to the canons whom he found in possession; yet he somehow bent

them all to his will, or carried their wills with his own.
'Never since I came to the diocese,' he said to his chaplain,
'have I had a quarrel with my chapter. It is not that I am
easy-going—sum enim reverâ pipere mordacior : pepper is not
more biting than I can be. I often fly out for small causes;
but they take me as they find me. There is not one who
distrusts my love for him, nor one by whom I do not believe
myself to be loved.'

At table this hardest of monks was the most agreeable of
companions. Though no one had practised abstinence more
severe, no one less valued it for its own sake, or had less
superstition or foolish sentiment about it. It was, and is,
considered sacrilege in the Church of Rome to taste food
before saying mass. Hugo, if he saw a priest who was to
officiate exhausted for want of support, and likely to find a
difficulty in getting through his work, would order him to eat
as a point of duty, and lectured him for want of faith if he
affected to be horrified.

Like all genuine men, the Bishop was an object of special
attraction to children and animals. The little ones in every
house that he entered were always found clinging about his
legs. Of the attachment of other creatures to him, there was
one very singular instance. About the time of his installation
there appeared on the mere at Stow Manor, eight miles from
Lincoln, a swan of unusual size, which drove the other male
birds from off the water. Abbot Adam, who frequently saw
the bird, says that he was curiously marked. The bill was
saffron instead of black, with a saffron tint on the plumage
of the head and neck; and the Abbot adds, he was as much
larger than other swans as a swan is larger than a goose.
This bird, on the occasion of the Bishop's first visit to the
manor, was brought to him to be seen as a curiosity. He
was usually unmanageable and savage; but the Bishop knew
the way to his heart; fed him, and taught him to poke his
head into the pockets of his frock to look for bread crumbs,
which he did not fail to find there. Ever after he seemed to
know instinctively when the Bishop was expected, flew
trumpeting up and down the lake, slapping the water with
his wings; when the horses approached, he would march

out upon the grass to meet them, strutted to the Bishop's side, and would sometimes follow him upstairs.

It was a miracle of course to the general mind, though explicable enough to those who have observed the physical charm which men who take pains to understand animals are able to exercise over them.

To relate, or even to sketch, Bishop Hugo's public life in the fourteen years that he was at Lincoln, would be beyond the compass of a magazine article. The materials indeed do not exist; for Abbot Adam's life is but a collection of anecdotes; and out of them it is only possible here to select a few at random. King Henry died two years after the scene at Woodstock; then came the accession of Cœur de Lion, the Crusade, the King's imprisonment in Austria, and the conspiracy of John. Glimpses can be caught of the Bishop in these stormy times quelling insurgent mobs—in Holland, perhaps Holland in Lincolnshire, with his brother William of Avalon, encountering a military insurrection; single-handed and unarmed, overawing a rising at Northampton, when the citizens took possession of the great church, and swords were flashing, and his attendant chaplains fled terrified, and hid themselves behind the altars.

These things, however, glad as we should be to know more of them, the Abbot merely hints at, confining himself to subjects more interesting to the convent recluses for whose edification he was writing.

But in whatever circumstances he lets us see the Bishop, it is always the same simple, brave, unpretending, wise figure, one to whom nature had been lavish of her fairest gifts, and whose training, to modern eyes so unpromising, had brought out all that was best in him.

Among the most deadly disorders which at that time prevailed in England was leprosy. The wretched creatures afflicted with so loathsome a disease were regarded with a superstitious terror : as the objects in some special way of the wrath of God. They were outlawed from the fellowship of mankind, and left to perish in misery.

The Bishop, who had clearer views of the nature and causes of human suffering, established hospitals on his estate for these poor victims of undeserved misery, whose misfortunes appeared to him to demand special care and sympathy. To the horror of his attendants, he persisted in visiting them himself; he washed their sores with his own hands, kissed them, prayed over them, and consoled them.

'Pardon, blessed Jesus,' explains Adam, 'the unhappy soul of him who tells the story! when I saw my master touch those bloated and livid faces; when I saw him kiss the bleared eyes or sightless sockets, I shuddered with disgust. But Hugo said to me that these afflicted ones were flowers of Paradise, pearls in the coronet of the Eternal King waiting for the coming of their Lord, who in His own time would change their forlorn bodies into the likeness of his own glory.'

He never altered his own monastic habits. He never parted with his hair shirt, or varied from the hardness of the Carthusian rule; but he refused to allow that it possessed any particular sanctity. Men of the world affected regret sometimes to him that they were held by duty to a secular life when they would have preferred to retire into a monastery. The kingdom of God, he used to answer, was not made up of monks and hermits. God, at the day of judgment, would not ask a man why he had not been a monk, but why he had not been a Christian. Charity in the heart, truth in the tongue, chastity in the body, were the virtues which God demanded : and chastity, to the astonishment of his clergy, he insisted, was to be found as well among the married as the unmarried. The wife was as honourable as the virgin. He allowed women (Adam's pen trembles as he records it) to sit at his side at dinner; and had been known to touch and even to embrace them. 'Woman,' he once said remarkably, 'has been admitted to a higher privilege than man. It has not been given to man to be the father of God. To woman it has been given to be God's mother.'

Another curious feature about him was his eagerness to be present, whenever possible, at the burial of the dead. He never allowed any one of his priests to bury a corpse if he were himself within reach. If a man had been good, he said, he deserved to be honoured. If he had been a sinner, there was the more reason to help him. He would allow nothing to interfere with a duty of this kind; and in great cities he would spend whole days by the side of graves. At Rouen once he was engaged to dinner with King Richard himself, and kept the King and the Court waiting for him while he was busy in the cemetery. A courtier came to fetch him. 'The King needn't wait,' he only said. 'Let him go to dinner in the name of God. Better the King dine without my company, than that I leave my Master's work undone.'

Gentle and affectionate as he shows himself in such traits as these, still, as he said, he was *pipere mordacior*—more biting than pepper. When there was occasion for anger there was fierce lightning in him; he was not afraid of the highest in the land.

The cause for which Becket died was no less dear to Hugo. On no pretext would he permit innovation on the Church's privileges, and he had many a sharp engagement with the primate, Archbishop Hubert, who was too complaisant to the secular power. An instance or two may be taken at random. There was a certain Richard de Wavre in his diocese, a younger son of a noble house, who was in deacon's orders, but the elder brother having died childless, was hoping to relapse into the lay estate. This Richard in some one of the many political quarrels of the day brought a charge of treason against Sir Reginald de Argentun, one of the Bishop's knights. As he was a clerk in orders the Bishop forbade him to appear as prosecutor in a secular court or cause. Cœur de Lion and Archbishop Hubert ordered him to go on. The Bishop suspended him for contumacy, the Archbishop removed the suspension. The Bishop pronounced sentence of excommunication; the Archbishop, as primate and legate, issued letters of absolution, which Richard flourished triumphantly in the Bishop's face.

'If my Lord Archbishop absolve you a hundred times,' was Hugo's answer, 'a hundred times I will excommunicate you again. Regard my judgment as you will, I hold you bound while you remain impenitent.' Death ended the dispute. The wretched Richard was murdered by one of his servants.

Another analogous exploit throws curious light on the habits of the times. Riding once through St Albans he met the sheriff with the *posse comitatus* escorting a felon to the gallows. The prisoner threw himself before the Bishop and claimed protection. The Bishop reined in his horse and asked who the man was.

'My Lord,' said the sheriff shortly, 'it is no affair of yours; let us pass and do our duty.'

'Eh!' then said Hugo. 'Blessed be God; we will see about that; make over the man to me; and go back and tell the judges that I have taken him from you.'

'My lords judges,' he said, when they came to remonstrate, 'I need not remind you of the Church's privilege of sanctuary; understand that where the Bishop is, the Church is. He who can consecrate the sanctuary carries with him the sacredness of the sanctuary.'

The humiliation of an English king at Becket's tomb had been a lesson too severe and too recent to be forgotten. 'We may not dispute you,' the judges replied; 'if you choose to let this man go we shall not oppose you, but you must answer for it to the King's Highness.'

'So be it,' answered Hugo, 'you have spoken well. I charge myself with your prisoner. The responsibility be mine.'

There was probably something more in the case than appears on the surface. The sanctuary system worked in mitigation of the law which in itself was frightfully cruel, and there may have been good reason why the life of the poor wretch should have been spared. The Bishop set him free. It is hoped that 'he sinned no more.'

The common-sense view which the Bishop took of miracles has been already spoken of, but we may give one or two other illustrations of it. Doubtless, he did not disbelieve in the possibility of miracles, but he knew how much imposture passed current under the name, and whether true or false he never missed a chance of checking or affronting superstition.

Stopping once in a country town on a journey from Paris to Troyes, he invited the parish priest to dine with him. The priest declined, but came in the evening to sit and talk with the chaplains. He was a lean old man, dry and shrivelled to the bones, and he told them a marvellous story which he bade them report to their master.

Long ago, he said, when he was first ordained priest, he fell into mortal sin, and without having confessed or done penance he had presumed to officiate at the altar. He was sceptical too, it seemed, a premature Voltairian. 'Is it credible,' he had said to himself when consecrating the host, 'that I, a miserable sinner, can manufacture and handle and eat the body and blood of God?' He was breaking the wafer at the moment; blood flowed at the fracture—the part which was in his hand became flesh. He dropped it terrified into the chalice, and the wine turned instantly into blood. The precious things were preserved. The priest went to Rome, confessed to the Pope himself, and received absolution. The faithful now flocked from all parts of France to adore the mysterious substances which were to be seen in the parish church; and the priest trusted that he might be honoured on the following day by the presence of Bishop Hugo and his retinue.

The chaplains rushed to the Bishop open-mouthed, eager to be allowed to refresh their souls on so divine a spectacle.

'In the name of God,' he said quietly, 'let unbelievers go rushing after signs and wonders. What have we to do with such things who partake every day of the heavenly sacrifice?' He dismissed the Priest with his blessing, giving him the benefit of a doubt, though he probably suspected him to be a rogue, and forbade his chaplains most strictly to yield to idle curiosity.

He was naturally extremely humorous, and humour in such men will show itself sometimes in playing with things, in the sacredness of which they may believe fully notwithstanding. It has been said, indeed, that no one has any real faith if he cannot afford to play with it.

Among the relics at Fécamp, in Normandy, was a so-called bone of Mary Magdalene. This precious jewel was kept with jealous care. It was deposited in a case, and within the case was double wrapped in silk. Bishop Hugo was taken to look at it in the presence of a crowd of monks, abbots, and other dignitaries; mass had been said first as a preparation; the thing was then taken out of its box and exhibited, so far it could be seen through its envelope. The Bishop asked to look at the bone itself; and no one venturing to touch it, he borrowed a knife and calmly slit the covering. He took it up, whatever it may have been, gazed at it, raised it to his lips as if to kiss it, and then suddenly with a strong grip of his teeth bit a morsel out of its side. A shriek of sacrilege rang through the church. Looking round quietly the Bishop said, ' Just now we were handling in our unworthy fingers the body of the Holy One of all. We passed Him between our teeth and down into our stomach; why may we not do the like with the members of his saints?'

We have left to the last the most curious of all the stories connected with this singular man. We have seen him with King Henry; we will now follow him into the presence of Cœur de Lion.

Richard, it will be remembered, on his return from his captivity plunged into war with Philip of France, carrying out a quarrel which had commenced in the Holy Land. The King, in distress for money, had played tricks with Church patronage which Hugo had firmly resisted. Afterwards an old claim on Lincoln diocese for some annual services was suddenly revived, which had been pretermitted for sixty years. The arrears for all that time were called for and exacted, and the clergy had to raise among themselves 3000 marks: hard measure of this kind perhaps induced Hugo to look closely into further demands.

In 1197, when Richard was in Normandy, a pressing message came home from him for supplies. A council was held at Oxford, when Archbishop Hubert, who was Chancellor, required each prelate and great nobleman in the King's name to provide three hundred knights at his own cost to serve in the war. The Bishop of London supported the primate. The Bishop of Lincoln followed. Being a stranger, he said, and ignorant on his arrival of English laws, he had made it his business to study them. The see of Lincoln, he was aware, was bound to military service, but it was service in England and not abroad. The demand of the King was against the liberties which he had sworn to defend, and he would rather die than betray them.

The Bishop of Salisbury, gathering courage from Hugo's

resistance, took the same side. The council broke up in confusion, and the Archbishop wrote to Richard to say that he was unable to raise the required force, and that the Bishop of Lincoln was the cause. Richard, who, with most noble qualities, had the temper of a fiend, replied instantly with an order to seize and confiscate the property of the rebellious prelates. The Bishop of Salisbury was brought upon his knees, but Hugo, fearless as ever, swore that he would excommunicate any man who dared to execute the King's command; and as it was known that he would keep his word, the royal officers hesitated to act. The King wrote a second time more fiercely, threatening death if they disobeyed, and the Bishop, not wishing to expose them to trouble on his account, determined to go over and encounter the tempest in person.

At Rouen, on his way to Roche d'Andeli, where Richard was lying, he was encountered by the Earl Marshal and Lord Albemarle, who implored him to send some conciliatory message by them, as the King was so furious that they feared he might provoke the anger of God by some violent act.

The Bishop declined their assistance. He desired them merely to tell the King that he was coming. They hurried back, and he followed at his leisure. The scene that ensued was even stranger than the interview described with Henry in the park at Woodstock.

Cœur de Lion, when he arrived at Roche d'Andeli, was hearing mass in the church. He was sitting in a great chair at the opening into the choir, with the Bishops of Durham and Ely on either side. Church ceremonials must have been conducted with less stiff propriety than at present. Hugo advanced calmly and made the usual obeisance. Richard said nothing, but frowned, looked sternly at him for a moment, and turned away.

'Kiss me, my Lord King,' said the Bishop. It was the ordinary greeting between the sovereign and the spiritual peers. The King averted his face still further.

'Kiss me, my Lord,' said Hugo again, and he caught Cœur de Lion by the vest and shook him, Abbot Adam standing shivering behind.

'Non meruisti—thou hast not deserved it,' growled Richard.

'I have deserved it,' replied Hugo, and shook him harder.

Had he shown fear, Cœur de Lion would probably have trampled on him, but who could resist such marvellous audacity? The kiss was given. The Bishop passed up to the altar and became absorbed in the service, Cœur de Lion curiously watching him.

When mass was over there was a formal audience, but the result of it was decided already. Hugo declared his loyalty in everything, save what touched his duty to God. The King yielded, and threw the blame of the quarrel on the too complaisant primate.

Even this was not all. The Bishop afterwards requested a private interview. He told Richard solemnly that he was uneasy for his soul, and admonished him, if he had anything on his conscience, to confess it.

The King said he was conscious of no sin, save of a certain rage against his French enemies.

'Obey God!' the Bishop said, 'and God will humble your enemies for you—and you for your part take heed you offend not Him or hurt your neighbour. I speak in sadness, but rumour says you are unfaithful to your queen.'

The lion was tamed for the moment. The King acknowledged nothing but restrained his passion, only observing afterwards, 'if all bishops were like my Lord of Lincoln, not a prince among us could lift his head against them.'

The trouble was not over. Hugo returned to England to find his diocese in confusion. A bailiff of the Earl of Leicester had taken a man out of sanctuary in Lincoln and had hung him. Instant excommunication followed. The Bishop compelled every one who had been concerned in the sacrilege to repair, stripped naked to the waist, to the spot where the body was buried, to dig it up, putrid as it was, and carry it on their shoulders round the town, to halt at each church door to be flogged by the priests belonging to the place, and then with their own hands to rebury the man in the cemetery from which he had been originally carried off.

Fresh demands for money in another, but no less irregular, form followed from the King. There was again a council in London. The Archbishop insisted that Hugo should levy a subsidy upon his clergy.

'Do you not know, my Lord,' the primate said, 'that the King is as thirsty for money as a man with the dropsy for water?'

'His Majesty may be dropsical for all that I know,' Hugo answered, 'but I will not be the water for him to swallow.'

Once more he started for Normandy, but not a second time to try the effect of his presence on Cœur de Lion. On approaching Angers he was met by Sir Gilbert de Lacy with the news that the Lion-heart was cold. Richard had been struck by an arrow in the trenches at Chaluz. The wound had mortified and he was dead. He was to be buried at Fontevrault, but the country was in the wildest confusion. The roads were patrolled by banditti, and de Lacy strongly advised the Bishop to proceed no further.

Hugo's estimate of danger was unlike de Lacy's. 'I have more fear,' he said, 'of failing through cowardice in my duty to my lord and prince. If the thieves take my horse and clothes from me, I can walk, and walk the lighter. If they tie me fast, I cannot help myself.'

Paying a brief visit to Queen Berengaria, at Beaufort Abbey, on the way, he reached Fontevrault on Palm Sunday, the day of the funeral, and was in time to pay the last honours to the sovereign whom he had defied and yet loved so dearly.

His own time was also nearly out, and this hurried sketch must also haste to its end. One more scene, however, remains to be described.

To Henry and Richard, notwithstanding their many faults, the Bishop was ardently attached. For their sakes, and for his country's, he did what lay in him to influence for good the brother who was to succeed to the throne.

At the time of Richard's death, John was with his nephew Arthur in Brittany. That John and not Arthur must take Richard's place the Bishop seems to have assumed as unavoidable; Arthur was but ten years old and the times were too rough for a regency. John made haste to Fontevrault, receiving on his way the allegiance of many of the barons. After the funeral he made a profusion of promises to the Bishop of Lincoln as to his future conduct.

The Bishop had no liking for John. He knew him to have been paltry, false, and selfish.

'I trust you mean what you say,' he said in reply. 'Nostis quia satis aversor mendacium,—you know that I hate lying.'

John produced an amulet which he wore round his neck with a chain. That he seemed to think would help him to walk straight.

The Bishop looked at it scornfully. 'Do you trust in a senseless stone?' he said. 'Trust in the living rock in heaven —the Lord Jesus Christ. Anchor your hopes in Him and He will direct you.'

On one side of the church at Fontevrault was a celebrated sculpture of the day of judgment. The Judge was on his throne; on his left were a group of crowned kings led away by devils to be hurled into the smoking pit. Hugo pointed significantly to them. 'Understand,' he said, 'that those men are going into unending torture. Think of it, and let your wisdom teach you the prospects of princes who, while they govern men, are unable to rule themselves, and become slaves in hell through eternity. Fear this, I say, while there is time. The hour will come when it will have been too late.'

John affected to smile, pointed to the good kings on the other side, and declared, with infinite volubility, that he would be found one of those.

The fool's nature, however, soon showed itself. Hugo took leave of him with a foreboding heart, paid one more bright brief visit in the following year to his birthplace in the south, and then returned to England to die. He had held his see but fourteen years, and was no more than sixty-five. His asceticism had not impaired his strength. At his last visit to the Chartreuse he had distanced all his companions on the steep hill-side, but illness overtook him on his way home. He arrived in London, at his house in the Old Temple, in the middle of September, to feel that he was rapidly dying. Of death itself, it is needless to say, he had no kind of fear. 'By the holy nut,' he used to say, in his queer way ('per sanctam nucem,[5] sic enim vice juramenti ad formationem verbi interdum loquebatur '), ' by the holy nut, we should be worse off if we were not allowed to die at all.'

He prepared with his unvarying composure. As his illness increased, and he was confined to his bed, his hair shirt hurt him. Twisting into knots, as he shifted from side to side, it bruised and wounded his skin. The rules of the order would have allowed him to dispense with it, but he could not be induced to let it go; but he took animal food, which the doctor prescribed as good for him, and quietly and kindly submitted to whatever else was ordered for him. He knew, however, that his life was over, and with constant confession held himself ready for the change. Great people came about him. John himself came; but he received him coldly. Archbishop Hubert came once; he did not care, perhaps, to return a second time.

The Archbishop, sitting by his bed, after the usual condolences, suggested that the Bishop of Lincoln might like to use the opportunity to repent of any sharp expressions which he had occasionally been betrayed into using. As the hint was not taken, he referred especially to himself as one of those who had something to complain of.

' Indeed, your Grace,' replied Hugo, ' there have been passages of words between us, and I have much to regret in relation to them. It is not, however, what I have said to your Grace, but what I have omitted to say. I have more feared to offend your Grace than to offend my Father in heaven. I have withheld words which I ought to have spoken, and I have thus sinned against your Grace and desire your forgiveness. Should it please God to spare my life I purpose to amend that fault.'

As his time drew near, he gave directions for the disposition of his body, named the place in Lincoln Cathedral where he was to be buried, and bade his chaplain make a cross of ashes on the floor of his room, lift him from his bed at the moment of departure, and place him upon it.

It was a November afternoon. The choristers of St Paul's were sent for to chant the compline to him for the last time.

[5] Perhaps for ' crucem,' as we say ' by *Gad,*' to avoid the actual word.

He gave a sign when they were half through. They lifted him and laid him on the ashes. The choristers sang on, and as they began the Nunc Dimittis he died.

So parted one of the most beautiful spirits that was ever incarnated in human clay. Never was man more widely mourned over, or more honoured in his death. He was taken down to Lincoln, and the highest and the lowest alike had poured out to meet the body. A company of Jews, the offscouring of mankind, for whom rack and gridiron were considered generally too easy couches, came to mourn over one whose justice had sheltered even them.

John was at Lincoln at the time, and William of Scotland with him; and on the hill, a mile from the town, two kings, three archbishops, fourteen bishops, a hundred abbots, and as many earls and barons, were waiting to receive the sad procession.

King John and the archbishops took the bier upon their shoulders, and waded knee-deep through the mud to the cathedral. The King of Scotland stood apart in tears.

It was no vain pomp or unmeaning ceremony, but the genuine healthful recognition of human worth. The story of Hugo of Lincoln has been too long unknown to us. It deserves a place in every biography of English Worthies. It ought to be familiar to every English boy. Such men as he were the true builders of our nation's greatness. Like the ' well-tempered mortar ' in old English walls, which is hard as the stone itself, their actions and their thoughts are the cement of our national organization, and bind together yet such parts of it as still are allowed to stand.

THE LIVES OF THE SAINTS[1]

If the enormous undertaking of the Bollandist editors had been completed, it would have contained the histories of 25,000 saints. So many the Catholic Church acknowledged and accepted as her ideals—as men who had not only done her honour by the eminence of their sanctity, but who had received while on earth an openly Divine recognition of it in gifts of supernatural power. And this vast number is but a selection; the editors chose only out of the mass before them what was most noteworthy and trustworthy, and what was of catholic rather than of national interest. It is no more than a fraction of that singular mythology which for so many ages delighted the Christian world, which is still held in external reverence among the Romanists, and of which the modern historians, provoked by its feeble supernaturalism, and by the entire absence of critical ability among its writers to distinguish between fact and fable, have hitherto failed to speak a reasonable word. Of the attempt in our own day to revive an interest in them we shall say little in this place. The ' Lives ' have no form or beauty to give them attraction in themselves; and for their human interest the broad atmosphere of the world suited ill with these delicate plants, which had grown up under the shadow of the convent wall; they were exotics, not from another climate, but from another age; the breath of scorn fell on them, and having no root in the hearts and beliefs of men any more, but only in the sentimentalities and make-beliefs, they withered and sank. And yet, in their place as historical phenomena, the legends of the saints are as remarkable as any of the Pagan mythologies; to the full as remarkable, perhaps far more so, if the length and firmness of hold they once possessed on the convictions of mankind is to pass for anything in the estimate—and to ourselves they have a near and peculiar interest, as spiritual facts in the growth of the Catholic faith.

[1] [1850]

Philosophy has rescued the old theogonies from ridicule; their extravagancies, even the most grotesque of them, can be now seen to have their root in an idea, often a deep one, representing features of natural history or of metaphysical speculation, and we do not laugh at them any more. In their origin, they were the consecration of the first-fruits of knowledge; the expression of a real reverential belief, and they could not grow; they became monstrous and mischievous, and were driven out by Christianity with scorn and indignation. But it is with human institutions as it is with men themselves; we are tender with the dead when their power to hurt us has passed away; and as Paganism can never more be dangerous, we have been able to command a calmer attitude towards it, and to detect under its most repulsive features sufficient latent elements of genuine thought to satisfy us that even in their darkest aberrations men are never wholly given over to falsehood and absurdity. When philosophy has done for mediæval mythology what it has done for Hesiod and for the Edda, we shall find there also at least a deep sense of the awfulness and mystery of life, and we shall find a moral element which the Pagans never had. The lives of the saints are always simple, often childish, seldom beautiful; yet, as Goethe observed, if without beauty, they are always good.

And as a phenomenon, let us not deceive ourselves on the magnitude of the Christian hagiology. The Bollandists were restricted on many sides. They took only what was in Latin—while every country in Europe had its own home growth in its own language—and thus many of the most characteristic of the lives are not to be found at all in their collection. And again, they took but one life of each saint, composed in all cases late, and compiled out of the mass of various shorter lives which had grown up in different localities out of popular tradition; so that many of their longer productions have an elaborate literary character, with an appearance of artifice, which, till we know how they came into existence, might blind us to the vast width and variety of the traditional sources from which they are drawn. In the twelfth century there were sixty-six lives extant of St Patrick alone; and that in a country where every parish had its own special saint and special legend of him. These sixty-six lives may have contained (Mr Gibbon says *must* have contained) at least as many thousand lies. Perhaps so. To severe criticism, even the existence of a single apostle, St Patrick, appears problematical. But at least there is the historical fact, about which there can be no mistake, that the stories did grow up in some way or other, that they were repeated, sung, listened to, written, and read; that these lives in Ireland, and all over Europe and over the earth wherever the Catholic faith was preached, stories like these, sprang out of the heart of the people, and grew and shadowed over the entire believing mind of the Catholic world. Wherever church was founded, or soil was consecrated for the long resting-place of those who had died in the faith; wherever the sweet bells of convent or of monastery were heard in the evening air, charming the unquiet world to rest and remembrance of God, there dwelt the memory of some apostle who had laid the first stone, there was the sepulchre of some martyr whose relics reposed beneath the altar, or some confessor who had suffered there for his Master's sake, of some holy ascetic who in silent self-chosen austerity had woven a ladder there of prayer and penance, on which the angels of God were believed to have ascended and descended. It is not a phenomenon of an age or of a century; it is characteristic of the history of Christianity. From the time when the first preachers of the faith passed out from their homes by that quiet Galilean lake, to go to and fro over the earth, and did their mighty work, and at last disappeared and were not any more seen, these sacred legends began to grow. Those who had once known the Apostles, who had drawn from their lips the blessed message of light and life, one and all would gather together what fragments they could find of their stories. Rumours blew in from all the winds. They had been seen here, had been seen there, in the farthest corners of the earth, preaching, contending, suffering, prevailing. Affection did not stay to scrutinize. When some member of a family among ourselves is absent in some far place from which sure news of him comes slowly and uncertainly; if he has been in the army, or on some dangerous expedition, or at sea, or anywhere where real or imaginary dangers stimulate anxiety; or when one is gone away from us altogether—fallen perhaps in battle —and when the story of his end can be collected but fitfully from strangers, who only knew his name, but had heard him nobly spoken of; the faintest threads are caught at; reports, the vagueness of which might be evident to indifference, are to love strong grounds of confidence, and ' trifles light as air ' establish themselves as certainties. So, in those first Christian communities, travellers came through from east and west; legions on the march, or caravans of wandering merchants; and one had been in Rome, and seen Peter disputing with Simon Magus; another in India, where he had heard St Thomas preaching to the Brahmins; a third brought with him, from the wilds of Britain, a staff which he had cut, as he said, from a thorn tree, the seed of which St Joseph had sown there, and which had grown to its full size in a single night, making merchandise of the precious relic out of the credulity of the believers. So the legends grew, and were treasured up, and loved, and trusted; and alas! all which we have been able to do with them is to call them lies, and to point a shallow moral on the impostures and credulities of the early Catholics. An Atheist could not wish us to say more. If we can really believe that the Christian Church was made over in its very cradle to lies and to the father of lies, and was allowed to remain in his keeping, so to say, till yesterday, he will not much trouble himself with any faith which after such an admission we may profess to entertain. For, as this spirit began in the first age in which the Church began to have a history, so it continued so long as the Church as an integral body retained its vitality, and only died out in the degeneracy which preceded and which brought on the Reformation. For fourteen hundred years these stories held their place and rang on from age to age, from century to century; as the new faith widened its boundaries, and numbered even more and more great names of men and women who had fought and died for it, so long their histories, living in the hearts of those for whom they laboured, laid hold of them and filled them; and the devout imagination, possessed with what was often no more than the rumour of a name, bodied it out into life, and form, and reality. And doubtless, if we try them by any historical canon, we have to say that quite endless untruths grew in this way to be believed among men; and not believed only, but held sacred, passionately and devotedly; not filling the history books only, not only serving to amuse and edify the refectory, or to furnish matter for meditation in the cell, but claiming days for themselves of special remembrance, entering into liturgies and inspiring prayers, forming the spiritual nucleus of the hopes and fears of millions of human souls.

From the hard barren standing ground of the fact idolator, what a strange sight must be that still mountain-peak on the wild west Irish shore, where, for more than ten centuries, a rude old bell and a carved chip of oak have witnessed, or seemed to witness, to the presence long ago there of the Irish apostle; and where, in the sharp crystals of the trap rock, a path has been worn smooth by the bare feet and bleeding knees of the pilgrims, who still, in the August weather, drag their painful way along it as they have done for a thousand years! Doubtless the 'Lives of the Saints' are full of lies. Are there none in the Iliad? or in the legends of Æneas? Were the stories sung in the liturgy of Eleusis all so true? so true as fact? Are the songs of the Cid or of Siegfried true? We say nothing of the lies in these; but why? Oh, it will be said, but they are fictions; they were never supposed to be true. But they *were* supposed to be true, to the full as true as the 'Legenda Aurea.'[2] Oh, then, they are poetry; and besides they have nothing to do with Christianity. Yes, that is it; they have nothing to do with Christianity. Religion has grown such a solemn business with us, and we bring such long faces to it, that we cannot admit or conceive to be at all naturally admissible such a light companion as the imagination. The distinction between secular and religious has been extended even to the faculties; and we cannot tolerate in others the fulness and freedom which we have lost or rejected for ourselves. Yet it has been a fatal mistake with the critics. They found themselves off the recognized ground of Romance and Paganism, and they failed to see the same principles at work, though at work with new materials. In the records of

[2] [A collection of lives of the saints compiled about the middle of the thirteenth century by the Dominican Jacobus De Voragine.]

all human affairs, it cannot be too often insisted on that two kinds of truth run ever side by side, or rather, crossing in and out with each other, form the warp and the woof of the coloured web which we call history; the one, the literal and external truths corresponding to the eternal and as yet undiscovered laws of fact; the other, the truths of feeling and of thought, which embody themselves either in distorted pictures of outward things, or in some entirely new creation— sometimes moulding and shaping real history; sometimes taking the form of heroic biography, of tradition, or popular legend; sometimes appearing as recognized fiction in the epic, the drama, or the novel. It is useless to tell us that this is to confuse truth and falsehood. We are stating a fact, not a theory; and if it makes truth and falsehood difficult to distinguish, that is nature's fault, not ours. Fiction is only false, when it is false, not to fact, else how could it be fiction? but when it is—to *law*. To try it by its correspondence to the real is pedantry. Imagination creates as nature creates, by the force which is in man, which refuses to be restrained; we cannot help it, and we are only false when we make monsters, or when we pretend that our inventions are facts, when we substitute truths of one kind for truths of another; when we stubstitute,—and again we must say when we *intentionally* substitute:—whenever persons and whenever facts seize strongly on the imagination (and of course when there is anything remarkable in them they must and will do so), invention glides into the images which form in our minds; so it must be, and so it ever has been, from the first legends of a cosmogony to the written life of the great man who died last year or century, or to the latest scientific magazine. We cannot relate facts as they are; they must first pass through ourselves, and we are more or less than mortal if they gather nothing in the transit. The great outlines alone lie around us as imperative and constraining; the detail we each fill up variously, according to the turn of our sympathies, the extent of our knowledge, or our general theories of things : and therefore it may be said that the only literally true history possible is the history which mind has left of itself in all the changes through which it has passed.

Suetonius is to the full as extravagant and superstitious as Surius,[3] and Suetonius was most laborious and careful, and was the friend of Tacitus and Pliny. Suetonius gives us prodigies, where Surius has miracles, but that is all the difference; each follows the form of the supernatural which belonged to the genius of his age. Plutarch writes a life of Lycurgus, with details of his childhood, and of the trials and vicissitudes of his age; and the existence of Lycurgus is now quite as questionable as that of St Patrick or of St George of England.

No rectitude of intention will save us from mistakes. Sympathies and antipathies are but synonyms of prejudice, and indifference is impossible. Love is blind, and so is every other passion. Love believes eagerly what it desires; it excuses or passes lightly over blemishes, it dwells on what is beautiful; while dislike sees a tarnish on what is brightest, and deepens faults into vices. Do we believe that all this is a disease of unenlightened times, and that in our strong sunlight only truth can get received?—then let us contrast the portrait, for instance, of Sir Robert Peel as it is drawn in the Free Trade Hall at Manchester,[4] at the county meeting, and in the Oxford Common Room. It is not so. Faithful and literal history is possible only to an impassive spirit. Man will never write it, until perfect knowledge and perfect faith in God shall enable him to see and endure every fact in its reality; until perfect love shall kindle in him under its touch the one just emotion which is in harmony with the eternal order of all things.

How far we are in these days from approximating to such a combination we need not here insist. Criticism in the hands of men like Niebuhr[5] seems to have accomplished great intellectual triumphs; and in Germany and France, and among ourselves, we have our new schools of the philosophy of history; yet their real successes have hitherto only been

[3] [L. Surius (1522-1578). A German Carthusian who published a collection of lives of the saints in 1570.]

[4] Written in 1850.

[5] [B. G. Niebuhr (1776-1831), founder of a more scientific school of historical criticism.]

destructive. When philosophy reconstructs, it does nothing but project its own idea; when it throws off tradition, it cannot work without a theory : and what is a theory but an imperfect generalization caught up by a predisposition? What is Comte's great division of the eras but a theory, and facts are but clay in his hands, which he can mould to illustrate it, as every clever man will find facts to be, let his theory be what it will? Intellect can destroy, but it cannot restore life; call in the creative faculties—call in Love, Idea, Imagination, and we have living figures, but we cannot tell whether they are figures which ever lived before. The high faith in which Love and Intellect can alone unite in their fulness, has not yet found utterance in modern historians.

The greatest man who has yet given himself to the recording of human affairs is, beyond question, Cornelius Tacitus. Alone in Tacitus a serene calmness of insight was compatible with intensity of feeling. He took no side; he may have been Imperialist, he may have been Republican, but he has left no sign whether he was either : he appears to have sifted facts with scrupulous integrity; to administer his love, his scorn, his hatred, according only to individual merit : and his sentiments are rather felt by the reader in the life-like clearness of his portraits, than expressed in words by himself. Yet such a power of seeing into things was only possible to him, because there was no party left with which he could determinedly side, and no wide spirit alive in Rome through which he could feel. The spirit of Rome, the spirit of life had gone away to seek other forms, and the world of Tacitus was a heap of decaying institutions; a stage where men and women, as they themselves were individually base or noble, played over their little parts. Life indeed was come into the world, was working in it, and silently shaping the old dead corpse into fresh and beautiful being. Tacitus alludes to it once only, in one brief scornful chapter; and the most poorly gifted of those forlorn biographers whose unreasoning credulity was piling up the legends of St Mary and the Apostles, which now drive the ecclesiastical historian to despair, knew more, in his divine hope and faith, of the real spirit which had gone out among mankind, than the keenest and gravest intellect which ever set itself to contemplate them.

And now having in some degree cleared the ground of difficulties, let us go back to the Lives of the Saints. If Bede tells us lies about St Cuthbert, we will disbelieve his stories; but we will not call Bede a liar, even though he prefaces his life with a declaration that he has set down nothing but what he has ascertained on the clearest evidence. We are driven to no such alternative; our canons of criticism are different from Bede's and so are our notions of probability. Bede would expect *à priori,* and would therefore consider as sufficiently attested by a consent of popular tradition, what the oaths of living witnesses would fail to make credible to a modern English jury. We will call Bede a liar only if he put forward his picture of St Cuthbert as a picture of a life which he considered admirable and excellent, as one after which he was endeavouring to model his own, and which he held up as a pattern of imitation, when in his heart he did not consider it admirable at all, when he was making no effort at the austerities which he was lauding. The histories of the saints are written as ideals of a Christian life; they have no elaborate and beautiful forms; single and straightforward as they are,—if they are not this they are nothing. For fourteen centuries the religious mind of the Catholic world threw them out as its form of hero worship, as the heroic patterns of a form of human life which each Christian within his own limits was endeavouring to realize. The first martyrs and confessors were to these poor monks what the first Dorian conquerors were in the war songs of Tyrtæus, what Achilles and Ajax and Agamemnon and Diomed were wherever Homer was sung or read; or in more modern times, what the Knights of the Round Table were in the halls of the Norman castles. The Catholic mind was expressing its conception of the highest human excellence; and the result is that immense and elaborate hagiology. As with the battle heroes, too, the inspiration lies in the universal idea; the varieties of character (with here and there an exception) are slight and unimportant; the object being to create examples for universal human imitation. Lancelot or Tristam were equally true to the spirit of chivalry; and Patrick on the mountain, or Antony in the desert, are equal models of patient austerity. The knights fight with giants, enchanters, robbers, unknightly nobles, or furious wild beasts; the Christians fight with the world, the flesh, and the devil. The knight leaves the comforts of home in quest of adventures, the saint in quest of penance, and on the bare rocks or in desolate wildernesses subdues the devil in his flesh with prayers and penances; and so alien is it all to the whole thought and system of the modern Christian, and he either rejects such stories altogether as monks' impostures, or receives them with disdainful wonder, as one more shameful form of superstition with which human nature has insulted heaven and disgraced itself.

Leaving, however, for the present, the meaning of monastic asceticism, it seems necessary to insist that there really was such a thing; there is no doubt about it. If the particular actions told of each saint are not literally true as belonging to him, abundance of men did for many centuries lead the sort of life which saints are said to have led. We have got a notion that the friars were a snug, comfortable set, after all; and the life in a monastery pretty much like that in a modern university, where the old monk's language and affectation of unworldliness does somehow contrive to co-exist with as large a mass of bodily enjoyment as man's nature can well appropriate. Very likely this was the state into which many of the monasteries had fallen in the fifteenth century. It was a symptom of a very rapid disorder which had set in among them, and which promptly terminated in dissolution. But long, long ages lay behind the fifteenth century, in which, wisely or foolishly, these old monks and hermits did make themselves a very hard life of it; and the legend only exceeded the reality in being a very slightly idealized portrait. We are not speaking of the miracles; that is a wholly different question. When men knew little of the order of nature, whatever came to pass without obvious cause was at once set down to influences beyond nature and above it; and so long as there were witches and enchanters, strong with the help of the bad powers, of course the especial servants of God would not be left without graces to outmatch and overcome the devil. And there were many other reasons why the saints should work miracles. They had done so under the old dispensation, and there was no obvious reason why Christians should be worse off than Jews. And again, although it be true, in the modern phrase, which is beginning to savour a little of cant, that the highest natural is the highest supernatural, nevertheless natural facts permit us to be so easily familiar with them, that they have an air of commonness; and when we have a vast idea to express, there is always a disposition to the extra-ordinary. But the miracles are not the chief thing; nor ever were they so. Men did not become saints by working miracles, but they worked miracles because they had become saints; and the instructiveness and value of their lives lay in the means which they had used to make themselves what they were; and as we said, in this part of the business there is unquestion-able basis of truth—scarcely even exaggeration. We have documentary evidence, which has been filtered through the sharp ordeal of party hatred, of the way in which some men (and those, not mere ignorant fanatics, but men of vast mind and vast influence in their days) conducted themselves, where *myth* has no room to enter. We know something of the hair-shirt of Thomas à Becket; and there was another poor monk, whose asceticism imagination could not easily outrun; he who, when the earth's mighty ones were banded together to crush him under their armed heels, spoke but one little word, and it fell among them like the spear of Cadmus; the strong ones turned their hands against each other, and the armies melted away; and the proudest monarch of the earth lay at that monk's threshold three winter nights in the scanty clothing of penance, suing miserably for forgiveness. Or again, to take a fairer figure. There is a poem extant, the genuineness of which, we believe, has not been challenged, composed by Columbkill, commonly called St Columba. He was a hermit in Arran, a rocky island in the Atlantic, outside Galway Bay; from which he was summoned, we do not know how, but in a manner which appeared to him to be a Divine call, to go away and be Bishop of Iona. The poem is a 'Farewell to Arran,' which he wrote on leaving it; and he lets us see something of a hermit's life there. 'Farewell,' he begins (we are obliged to quote from memory), 'a long farewell to thee, Arran of my heart. Paradise is with thee; the garden of God within the sound of thy bells. The angels love Arran. Each day an angel comes there to join in its services.' And then he goes on to describe his 'dear cell,' and the holy happy hours which he had spent there, 'with the wind whistling through the loose stones, and the sea spray hanging on his hair.' Arran is no better than a wild rock. It is strewed over with the ruins which may still be seen of the old hermitages; and at their best they could have been but such places as sheep would huddle under in a storm, and shiver in the cold and wet which would pierce through the chinks of the walls.

Or, if written evidence be too untrustworthy, there are silent witnesses which cannot lie, that tell the same touching story. Whoever loiters among the ruins of a monastery will see, commonly leading out of the cloisters, rows of cellars half under-ground, low, damp, and wretched-looking; an earthen floor, bearing no trace of pavement; a roof from which the mortar and the damp keep up (and always must have kept up) a perpetual ooze; for a window a narrow slip in the wall, through which the cold and the wind find as free an access as the light. Such as they are, a well-kept dog would object to accept a night's lodging in them; and if they had been prison cells, thousands of philanthropic tongues would have trumpeted out their horrors. The stranger perhaps supposes that they were the very dungeons of which he has heard such terrible things. He asks his guide, and his guide tells him they were the monks' dormitories. Yes; there on that wet soil, with that dripping roof above them, was the self-chosen home of those poor men. Through winter frost, through rain and storm, through summer sunshine, generation after generation of them, there they lived and prayed, and at last lay down and died.

It is all gone now—gone as if it had never been; and it was as foolish as, if the attempt had succeeded, it would have been mischievous, to revive a devotional interest in the Lives of the Saints. It would have produced but one more unreality in an age already too full of such. No one supposes we should have set to work to live as they lived; that any man, however earnest in his religion, would have gone looking for earth floors and wet dungeons, or wild islands to live in, when he could get anything better. Either we are wiser, or more humane, or more self-indulgent; at any rate we are something which divides us from mediæval Christianity by an impassible gulf which this age or this epoch will not see bridged over. Nevertheless, these modern hagiologists, however wrongly they went to work at it, had detected, and were endeavouring to fill, a very serious blank in our educational system; a very serious blank indeed, and one which, somehow, we must contrive to get filled if the education of character is ever to be more than a name to us. To try and teach people how to live without giving them examples in which our rules are illustrated, is like teaching them to draw by the rules of perspective, and of light and shade, without designs in which to study the effects; or to write verse by the laws of rhyme and metre, without song or poem in which rhyme and metre are exhibited. It is a principle which we have forgotten, and it is one which the old Catholics did not forget. We do not mean that they set out with saying to themselves, ' we must have examples, we must have ideals;' very likely they never thought about it at all; love for their holy men, and a thirst to know about them, produced the histories, and love unconsciously working gave them the best for which they could have wished. The boy at school at the monastery, the young monk disciplining himself as yet with difficulty under the austerities to which he had devoted himself, the old one halting on toward the close of his pilgrimage,—all of them had before their eyes, in the legend of the patron saint, a personal realization of all they were trying after; leading them on, beckoning to them, and pointing, as they stumbled among their difficulties, to the marks which his own footsteps had left, as he had trod that hard path before them. It was as if the Church was for ever saying to them : ' You have doubts and fears, and trials and temptations, outward and inward; you have sinned, perhaps, and feel the burden of your sin. Here was one who, like you, *in this very spot*, under the same sky, treading the same soil, among the same hills and woods and rocks and rivers, was tried like you, tempted like you, sinned like you; but here he prayed, and persevered, and did penance, and washed out his sins; he fought the fight, he vanquished the Evil One, he triumphed, and now he reigns with Christ in heaven. The same ground which yields you your food, once supplied him; he breathed, and lived, and felt, and died *here;* and now, from his throne in the sky, he is still looking lovingly down on his children, making intercession for you that you may have grace to follow him, that by-and-by he may himself offer you at God's throne as his own.' It is impossible to measure the influence which a personal reality of this kind must have exercised on the mind, thus daily and hourly impressed upon it through a life; there is nothing vague any more, no abstract excellences to strain after; all is distinct, personal, palpable. It is no dream. The saint's bones are under the altar; nay, perhaps, his very form and features undissolved. Under some late abbot the coffin may have been opened and the body seen without mark or taint of decay. Such things have been, and the emaciation of a saint will account for it without a miracle. Daily some incident of his story is read aloud, or spoken of, or preached upon. In quaint beautiful forms it lives in light in the long chapel windows; and in the summer matins his figure, lighted up in splendour, gleams down on the congregation as they pray, or streams in mysterious tints along the pavement, clad, as it seems, in soft celestial glory, and shining as he shines in heaven. Alas, alas! where is it all gone?

We are going to venture a few thoughts on the wide question, what possibly may have been the meaning of so large a portion of the human race, and so many centuries of Christianity, having been surrendered and seemingly sacrificed to the working out this dreary asceticism. If right once, then it is right now; if now worthless, then it could never have been more than worthless; and the energies which spent themselves on it were like corn sown upon the rock, or substance given for that which is not bread. We supposed ourselves challenged recently for our facts. Here is an enormous fact which there is no evading. It is not to be slurred over with indolent generalities, with unmeaning talk of superstition, of the twilight of the understanding, of barbarism, and of nursery credulity; it is matter for the philosophy of history, if the philosophy has yet been born which can deal with it; one of the solid, experienced facts in the story of mankind which must be accepted and considered with that respectful deference which all facts claim of their several sciences, and which will certainly not disclose its meaning (supposing it to have a meaning) except to reverence, to sympathy, to love. We must remember that the men who wrote these stories, and who practised these austerities, were the same men who composed our liturgies, who built our churches and our cathedrals—and the gothic cathedral is, perhaps, on the whole, the most magnificent creation which the mind of man has as yet thrown out of itself. If there be any such thing as a philosophy of history, real or possible, it is in virtue of there being certain progressive organizing laws in which the fretful lives of each of us are gathered into and subordinated in some larger unity, through which age is linked to age, as we move forward, with an horizon expanding and advancing. And if this is true, the magnitude of any human phenomenon is a criterion of its importance, and definite forms of thought working through long historic periods imply an effect of one of these vast laws—imply a distinct step in human progress. Something previously unrealized is being lived out, and rooted into the heart of mankind.

Nature never half does her work. She goes over it, and over it, to make assurance sure, and makes good her ground with wearying repetition. A single section of a short paper is but a small space to enter on so vast an enterprise; nevertheless, a few very general words shall be ventured as a suggestion of what this monastic or saintly spirit may possibly have meant.

First, as the spirit of Christianity is antagonistic to the world, whatever form the spirit of the world assumes, the ideals of Christianity will of course be their opposite; as one verges into one extreme, the other will verge into the contrary. In those rough times the law was the sword; animal might of arm, and the strong animal heart which guided it, were the excellences which the world rewarded; and monasticism, therefore, in its position of protest, would be the destruction and abnegation of the animal nature. The war hero in the battle or the tourney yard might be taken as the apotheosis of the fleshly man—the saint in the desert of the spiritual.

But this interpretation is slight, imperfect, and if true at all only partially so. The animal and the spiritual are not contradictories; they are the complements in the perfect character; and in the middle ages, as all ages of genuine earnestness, they interfused and penetrated each other. There were warrior saints, and saintly warriors; and those grand old figures which sleep cross-legged in the cathedral aisles were something higher than only one more form of the beast of prey. Monasticism represented something more positive than a protest against the world. We believe it to have been the realization of the infinite loveliness and beauty of personal purity.

In the earlier civilization, the Greeks, however genuine their reverence for the gods, do not seem to have supposed any part of their duty to the gods to consist in keeping their bodies untainted. Exquisite as was their sense of beauty, of beauty of mind as well as beauty of form, with all their loftiness and their nobleness, with their ready love of moral excellence when manifested, as fortitude, or devotion to liberty and to home, they had little or no idea of what we mean by morality. With a few rare exceptions, pollution, too detestable to be even named among ourselves, was of familiar and daily occurrence among their greatest men; was no reproach to philosopher or to statesman; and was not supposed to be incompatible, and was not, in fact, incompatible with any of those especial excellences which we so admire in the Greek character.

Among the Romans (that is, the early Romans of the Republic), there was a sufficiently austere morality. A public officer of state, whose business was to inquire into the private lives of the citizens, and to punish offences against morals, is a phenomenon which we have seen only once on this planet. There was never a nation before, and there has been none since, with sufficient virtue to endure it. But the Roman morality was not lovely for its own sake, nor excellent in itself. It was obedience to law, practised and valued, loved for what resulted from it, for the strength and rigid endurance which it gave, but not loved for itself. The Roman nature was fierce, rugged, almost brutal; and it submitted to restraint as stern as itself, as long as the energy of the old spirit endured. But as soon as that energy grew slack—when the religion was no longer believed, and taste, as it was called, came in, and there was no more danger to face, and the world was at their feet, all was swept away as before a whirlwind; there was no loveliness in virtue to make it desired, and the Rome of the Cæsars presents, in its later ages, a picture of enormous sensuality, of the coarsest animal desire, with means unlimited to gratify it. In Latin literature, as little as in the Greek, is there any sense of the beauty of purity. Moral essays on temperance we may find, and praise enough of the wise men whose passions and whose appetites are trained into obedience to reason. But this is no more than the philosophy of the old Roman life, which got itself expressed in words when men were tired of the reality. It involves no sense of sin. If sin could be indulged without weakening self-command, or without hurting other people, Roman philosophy would have nothing to say against it.

The Christians stepped far out beyond philosophy. Without speculating on the *why*, they felt that indulgence of animal passion did, in fact, pollute them, and so much the more, the more it was deliberate. Philosophy, gliding into Manicheism,[6] divided the forces of the universe, giving the spirit to God, but declaring matter to be eternally and incurably evil; and looking forward to the time when the spirit should be emancipated from the body, as the beginning of, or as the return to, its proper existence, a man like Plotinus[7] took no especial care what became the meanwhile of its evil tenement of flesh. If the body sinned, sin was its element; it could not do other than sin; purity of conduct could not make the body clean, and no amount of bodily indulgence could shed a taint upon the spirit—a very comfortable doctrine, and one which,

---

[6] [Named from the Syrian Mani (215-276 A.D.), whose views were based on the dualism of good and evil.]

[7] [Born in Egypt 204 A.D. Founder of the Neo-Platonic philosophy.]

under various disguises, has appeared a good many times on the earth. But Christianity, shaking all this off, would present the body of God as a pure and holy sacrifice, as so much of the material world conquered from the appetites and lusts, and from the devil whose abode they were. This was the meaning of the fastings and scourgings, the penances and night-watchings; it was this which sent St Anthony to the tombs and set Simeon on his pillar, to conquer the devil in the flesh, and keep themselves, if possible, undefiled by so much as one corrupt thought.

And they may have been absurd and extravagant. When the feeling is stronger than the judgment, men are very apt to be extravagant. If, in the recoil from Manicheism, they conceived that a body of a saint thus purified had contracted supernatural virtue and could work miracles, they had not sufficiently attended to the facts, and so far are not unexceptionable witnesses to them. Nevertheless they did their work, and in virtue of it we are raised to a higher stage—we are lifted forward a mighty step which we can never again retrace. Personal purity is not the whole for which we have to care : it is but one feature in the ideal character of man. The monks may have thought it was all, or more nearly all than it is; and therefore their lives may seem to us poor, mean, and emasculate. Yet it is with life as it is with science; generations of men have given themselves exclusively to single branches, which, when mastered, form but a little section in a cosmic philosophy; and in life, so slow is progress, it may take a thousand years to make good a single step. Weary and tedious enough it seems when we cease to speak in large language, and remember the numbers of individual souls who have been at work at the process; but who knows whereabouts we are in the duration of the race? Is humanity crawling out of the cradle, or tottering into the grave? Is it in nursery, in schoolroom, or in opening manhood? Who knows? It is enough for us to be sure of our steps when we have taken them, and thankfully to accept what has been done for us. Henceforth it is impossible for us to give our unmixed admiration to any character which moral shadows overhang. Henceforth we require, not greatness only, but goodness; and not that goodness only which begins and ends in conduct correctly regulated, but that love of goodness, that keen pure feeling for it, which resides in a conscience as sensitive and susceptible as a woman's modesty.

So much for what seems to us the philosophy of this matter. If we are right, it is no more than a first furrow in the crust of a soil which hitherto the historians have been contented to leave in its barrenness. If they are conscientious enough not to trifle with the facts, as they look back on them from the luxurious self-indulgence of modern Christianity, they either revile the superstition or pity the ignorance which made such large mistakes on the nature of religion—and, loud in their denunciations of priestcraft and of lying wonders, they point their moral with pictures of the ambition of mediæval prelacy or the scandals of the annals of the Papacy. For the inner life of all those millions of immoral souls who were struggling, with such good or bad success as was given them, to carry Christ's cross along their journey through life, they set it by, pass it over, dismiss it out of history, with some poor commonplace simper of sorrow or of scorn. It will not do. Mankind have not been so long on this planet altogether, that we can allow so large a chasm to be scooped out of their spiritual existence.

We intended to leave our readers with something lighter than all this in the shape of literary criticism, and a few specimens of the biographical style; in both of these we must now, however, be necessarily brief. Whoever is curious to study the lives of the saints in their originals, should rather go anywhere than to the Bollandists, and universally never read a late life when he can command an early one; for the genius in them is in the ratio of their antiquity, and, like river-water, is most pure nearest to the fountain. We are lucky in possessing several specimens of the mode of their growth in late and early lives of the same saints, and the process in all is similar. Out of the unnumbered lives of St Bride, three are left; out of the sixty-six of St Patrick, there are eight; the first of each belonging to the sixth century, the latest to the thirteenth. The earliest in each instance are in verse; they belong to a time when there was no one to write such things, and were popular in form and popular in their origin. The flow is easy, the style graceful and natural; but the step from poetry to prose is substantial as well as formal; the imagination is ossified, and we exchange the exuberance of legendary creativeness for the dogmatic record of fact without reality, and fiction without grace. The marvellous in the poetical lives is comparatively slight; the after-miracles being composed frequently out of a mistake of poets' metaphors for literal truth. There is often real, genial, human beauty in the old verse. The first two stanzas, for instance, of St Bride's Hymn are of high merit, as may, perhaps, be imperfectly seen in a translation:—

> *Bride the queen, she loved not the world;*
> *She floated on the waves of the world*
> *As the sea-bird floats upon the billow.*
> *Such sleep she slept as the mother sleeps*
> *In the far land of her captivity,*
> *Mourning for her child at home.*

What a picture is there of the strangeness and yearning of the poor human soul in this earthly pilgrimage!

The poetical 'Life of St Patrick,' too, is full of fine, wild, natural imagery. The boy is described as a shepherd on the hills of Down, and there is a legend, well told, of the angel Victor coming to him, and leaving a gigantic foot-print on a rock from which he sprang back into heaven. The legend, of course, rose from some remarkable natural feature of the spot; as it is first told, a shadowy unreality hangs over it, and it is doubtful whether it is more than a vision of the boy; but in the later prose all is crystalline; the story is drawn out, with a barren prolixity of detail, into a series of angelic visitations. And again, when Patrick is described, as the after-apostle, raising the dead Celts to life, the metaphor cannot be left in its natural force, and we have a long weary list of literal deaths and literal raisings. So in many ways the freshness and individuality was lost with time. The larger saints swallowed up the smaller and appropriated their exploits; chasms were supplied by an ever-ready fancy; and, like the stock of good works laid up for general use, there was a stock of miracles ever ready when any defect was to be supplied. So it was that, after the first impulse, the progressive life of a saint rolled on like a snowball down a mountain-side, gathering up into itself whatever lay in its path, fact or legend, appropriate or inappropriate—sometimes real jewels of genuine old tradition, sometimes the débris of the old creeds and legends of heathenism; and on, and on, till at length it reached the bottom, and was dashed in pieces on the Reformation.

One more illustration shall serve as evidence of what the really greatest, most vigorous, minds in the twelfth century could accept as possible or probable, which they could relate (on what evidence we do not know) as really ascertained facts. We remember something of St Anselm : both as a statesman and as a theologian, he was unquestionably among the ablest men of his time alive in Europe. Here is a story which Anselm tells of a certain Cornish St Kieran. The saint, with thirty of his companions, was preaching within the frontiers of a lawless Pagan prince; and, disregarding all orders to be quiet or to leave the country, continued to agitate, to threaten, and to thunder even in the ears of the prince himself. Things took their natural course. Disobedience provoked punishment. A guard of soldiers was sent, and the saint and his little band were decapitated. The scene of the execution was a wood, and the heads and trunks were left lying there for the wolves and the wild birds.

> But now a miracle, such as was once heard of before in the Church in the person of the holy Denis, was again wrought by Divine Providence to preserve the bodies of these saints from profanation. The trunk of Kieran rose from the ground, and selecting first his own head, and carrying it to a stream, and there carefully washing it, and afterwards performing the same sacred office for each of his companions, giving each body its own head, he dug graves for them and buried them, and last of all buried himself.

It is even so. So it stands written in a life claiming Anselm's authorship; and there is no reason why the authorship should not be his. Out of the heart come the issues of evil and of good, and not out of the intellect or the understanding.

Men are not good or bad, noble or base—thank God for it!—as they judge well or ill of the probabilities of nature, but as they love God and hate the devil. And yet the story is instructive. We have heard grave good men—men of intellect and influence—with all the advantages of modern science, learning, experience; men who would regard Anselm with sad and serious pity; yet tell us stories, as having fallen within their own experience, of the marvels of mesmerism, to the full as ridiculous (if anything is ridiculous) as this of the poor decapitated Kieran.

Mutato nomine, de te
Fabula narratur.

We see our natural faces in the glass of history, and turn away and straightway forget what manner of men we are. The superstition of science scoffs at the superstition of faith.

THE DISSOLUTION OF THE
MONASTERIES[1]

To be entirely just in our estimate of other ages is not difficult
—it is impossible. Even what is passing in our presence we
see but through a glass darkly. The mind as well as the eye
adds something of its own, before an image, even of the
clearest object, can be painted upon it.

And in historical inquiries, the most instructed thinkers
have but a limited advantage over the most illiterate. Those
who know the most, approach least to agreement. The most
careful investigations are diverging roads—the further men
travel upon them, the greater the interval by which they are
divided. In the eyes of David Hume the history of the Saxon
Princes is ' the scuffling of kites and crows.' Father Newman
would mortify the conceit of a degenerate England by pointing
to the sixty saints and the hundred confessors who were trained
in her royal palaces for the Calendar of the Blessed. How vast
a chasm yawns between these two conceptions of the same
era! Through what common term can the student pass from
one into the other?

Or, to take an instance yet more noticeable. The history
of England scarcely interests Mr Macaulay before the Revolution of the seventeenth century. To Lord John Russell, the
Reformation was the first outcome from centuries of folly and
ferocity; and Mr Hallam's more temperate language softens,
without concealing, a similar conclusion. These writers have
all studied what they describe. Mr Carlyle has studied the
same subject with power at least equal to theirs, and to him
the greatness of English character was waning with the dawn
of English literature; the race of heroes was already failing.
The era of action was yielding before the era of speech.

All these views may seem to ourselves exaggerated; we
may have settled into some moderate *via media*, or have carved
out our own ground on an original pattern; but if we are

[1] From *Fraser's Magazine*, 1857.

wise, the differences in other men's judgments will teach us to be diffident. The more distinctly we have made history bear witness in favour of our particular opinions, the more we have multiplied the chances against the truth of our own theory.

Again, supposing that we have made a truce with ' opinions,' properly so called; supposing we have satisfied ourselves that it is idle to quarrel upon points on which good men differ, and that it is better to attend rather to what we certainly know; supposing that, either from superior wisdom, or from the conceit of superior wisdom, we have resolved that we will look for human perfection neither exclusively in the Old World nor exclusively in the New—neither among Catholics nor Protestants, among Whigs or Tories, heathens or Christians—that we have laid aside accidental differences, and determined to recognize only moral distinctions, to love moral worth, and to hate moral evil, wherever we find them; —even supposing all this, we have not much improved our position—we cannot leap from our shadow.

Eras, like individuals, differ from one another in the species of virtue which they encourage. In one age, we find the virtues of the warrior; in the next, of the saint. The ascetic and the soldier in their turn disappear; an industrial era succeeds, bringing with it the virtues of common sense, of grace, and refinement. There is the virtue of energy and command, there is the virtue of humility and patient suffering. All these are different, and all are, or may be, of equal moral value; yet, from the constitution of our minds, we are so framed that we cannot equally appreciate all; we sympathize instinctively with the person who most represents our own ideal—with the period when the graces which most harmonize with our own tempers have been especially cultivated. Further, if we leave out of sight these refinements, and content ourselves with the most popular conceptions of morality, there is this immeasurable difficulty—so great, yet so little considered,— that goodness is positive as well as negative, and consists in the active accomplishment of certain things which we are bound to do, as well as in the abstaining from things which we are bound not to do. And here the warp and woof vary in shade and pattern. Many a man, with the help of circumstances, may pick his way clear through life, having never violated one prohibitive commandment, and yet at last be fit only for the place of the unprofitable servant—he may not have committed either sin or crime, yet never have felt the pulsation of a single unselfish emotion. Another, meanwhile, shall have been hurried by an impulsive nature into fault after fault—shall have been reckless, improvident, perhaps profligate, yet be fitter after all for the kingdom of heaven than the Pharisee— fitter, because against the catalogue of faults there could perhaps be set a fairer list of acts of comparative generosity and self-forgetfulness—fitter, because to those who love much, much is forgiven. Fielding had no occasion to make Blifil, behind his decent coat, a traitor and a hypocrite. It would have been enough to have coloured him in and out alike in the steady hues of selfishness, afraid of offending the upper powers as he was afraid of offending Allworthy—not from any love for what was good, but solely because it would be imprudent—because the pleasure to be gained was not worth the risk of consequences. Such a Blifil would have answered the novelist's purpose—for he would have remained a worse man in the estimation of some of us than Tom Jones.

So the truth is; but unfortunately it is only where accurate knowledge is stimulated by affection, that we are able to feel it. Persons who live beyond our own circle, and, still more, persons who have lived in another age, receive what is called justice, not charity; and justice is supposed to consist in due allotments of censure for each special act of misconduct, leaving merit unrecognized. There are many reasons for this harsh method of judging. We must decide of men by what we know, and it is easier to know faults than to know virtues. Faults are specific, easily described, easily appreciated, easily remembered. And again, there is, or may be, hypocrisy in virtue; but no one pretends to vice who is not vicious. The bad things which can be proved of a man we know to be genuine. He was a spendthrift, he was an adulterer, he gambled, he equivocated. These are blots positive, unless untrue, and when they stand alone, tinge the whole character.

This also is to be observed in historical criticism. All men feel a necessity of being on some terms with their conscience, at their own expense or at another's. If they cannot part with their faults, they will at least call them by their right name when they meet with such faults elsewhere; and thus, when they find account of deeds of violence or sensuality, of tyranny, of injustice of man to man, of great and extensive suffering, or any of those other misfortunes which the selfishness of men has at various times occasioned, they will vituperate the doers of such things, and the age which has permitted them to be done, with the full emphasis of virtuous indignation, while all the time they are themselves doing things which will be described, with no less justice, in the same colours, by an equally virtuous posterity.

Historians are fond of recording the supposed sufferings of the poor in the days of serfdom and villenage; yet the records of the strikes of the last ten years, when told by the sufferers, contain pictures no less fertile in tragedy. We speak of famines and plagues under the Tudors and Stuarts; but the Irish famine, and the Irish plague of 1847, the last page of such horrors which has yet been turned over, is the most horrible of all. We can conceive a description of England during the year which has just closed over us (1856), true in all its details, containing no one statement which can be challenged, no single exaggeration which can be proved; and this description, if given without the correcting traits, shall make ages to come marvel why the Cities of the Plain were destroyed, and England was allowed to survive. The frauds of trusted men, high in power and high in supposed religion; the wholesale poisonings; the robberies; the adulteration of food—nay, of almost everything exposed for sale—the cruel usage of women—children murdered for the burial fees—life and property insecure in open day in the open streets—splendour such as the world never saw before upon earth, with vice and squalor crouching under its walls—let all this be written down by an enemy, or let it be ascertained hereafter by the investigation of a posterity which desires to judge us as we generally have judged our forefathers, and few years will show darker in the English annals than the year which we have just left behind us. Yet we know, in the honesty of our hearts, how unjust such a picture would be. Our future advocate, if we are so happy as to find one, may not be able to disprove a single article in the indictment; and yet we know that, as the world goes, he will be right if he marks the year with a white stroke—as one of which, on the whole, the moral harvest was better than an average.

Once more : our knowledge of any man is always inadequate —even of the unit which each of us calls himself; and the first condition under which we can know a man at all is, that he be in essentials something like ourselves; that our own experience be an interpreter which shall open the secrets of his experience; and it often happens, even among our contemporaries, that we are altogether baffled. The Englishman and the Italian may understand each other's speech, but the language of each other's ideas has still to be learnt. Our long failures in Ireland have risen from a radical incongruity of character which has divided the Celt from the Saxon. And again, in the same country, the Catholic will be a mystery to the Protestant, and the Protestant to the Catholic. Their intellects have been shaped in opposite moulds; they are like instruments which cannot be played in concert. In the same way, but in a far higher degree, we are divided from the generations which have preceded us in this planet—we try to comprehend a Pericles or a Cæsar—an image rises before us which we seem to recognize as belonging to our common humanity. There is this feature which is familiar to us— and this—and this. We are full of hope; the lineaments, one by one, pass into clearness; when suddenly the figure becomes enveloped in a cloud—some perplexity crosses our analysis, baffling it utterly, the phantom which we have evoked dies away before our eyes, scornfully mocking our incapacity to master it.

The English antecedents to the Reformation are nearer to us than Greeks or Romans; and yet there is a large interval between the baron who fought at Barnet field, and his polished descendant in a modern drawing-room. The scale of appreciation and the rule of judgment—the habits, the hopes, the fears, the emotions—have utterly changed.

In perusing modern histories, the present writer has been struck dumb with wonder at the facility with which men will fill in chasms in their information with conjecture; will guess at the motives which have prompted actions; will pass their censures, as if all secrets of the past lay out on an open scroll before them. He is obliged to say for himself that, wherever he has been fortunate enough to discover authentic explanations of English historical difficulties, it is rare indeed that he has found any conjecture, either of his own or of any other modern writer, confirmed. The true motive has almost invariably been of a kind which no modern experience could have suggested.

Thoughts such as these form a hesitating prelude to an expression of opinion on a controverted question. They will serve, however, to indicate the limits within which the said opinion is supposed to be hazarded. And, in fact, neither in this nor in any historical subject is the conclusion so clear that it can be enunciated in a definite form. The utmost which can be safely hazarded with history is to relate honestly ascertained facts, with only such indications of a judicial sentence upon them as may be suggested in the form in which the story is arranged.

Whether the monastic bodies of England, at the time of their dissolution, were really in that condition of moral corruption which is laid to their charge in the Act of Parliament by which they were dissolved, is a point which it seems hopeless to argue. Roman Catholic, and indeed almost all English, writers who are not committed to an unfavourable opinion by the ultra-Protestantism of their doctrines, seem to have agreed of late years that the accusations, if not false, were enormously exaggerated. The dissolution, we are told, was a predetermined act of violence and rapacity; and when the reports and the letters of the visitors are quoted in justification of the Government, the discussion is closed with the dismissal of every unfavourable witness from the court, as venal, corrupt, calumnious—in fact, as a suborned liar. Upon these terms the argument is easily disposed of; and if it were not that truth is in all matters better than falsehood, it would be idle to reopen a question which cannot be justly dealt with.

No evidence can affect convictions which have been arrived at without evidence—and why should we attempt a task which is hopeless to accomplish? It seems necessary, however, to reassert the actual state of the surviving testimony from time to time, if it be only to sustain the links of the old traditions; and the present paper will contain one or two pictures of a peculiar kind, exhibiting the life and habits of those institutions, which have been lately met with chiefly among the unprinted Records. In anticipation of any possible charge of unfairness in judging from isolated instances, we disclaim simply all desire to judge—all wish to do anything beyond relating certain ascertained stories. Let it remain, to those who are perverse enough to insist upon it, an open question whether the monasteries were more corrupt under Henry the Eighth than they had been four hundred years earlier. The dissolution would have been equally a necessity; for no reasonable person would desire that bodies of men should have been maintained for the only business of singing masses, when the efficacy of masses was no longer believed. Our present desire is merely this—to satisfy ourselves whether the Government, in discharging a duty which could not be dispensed with, condescended to falsehood in seeking a vindication for themselves which they did not require; or whether they had cause really to believe the majority of the monastic bodies to be as they affirmed—whether, that is to say, there really were such cases either of flagrant immorality, neglect of discipline, or careless waste and prodigality, as to justify the general censure which was pronounced against the system by the Parliament and the Privy Council.

Secure in the supposed completeness with which Queen Mary's agents destroyed the Records of the visitation under her father, Roman Catholic writers have taken refuge in a disdainful denial; and the Anglicans, who for the most part, while contented to enjoy the fruits of the Reformation, detest the means by which it was brought about, have taken the same view. Bishop Latimer tells us that, when the Report of the visitors of the abbeys was read in the Commons House, there rose from all sides one long cry of ' Down with them.' But Bishop Latimer, in the opinion of High Churchmen, is not to be believed. Do we produce letters of the visitors themselves, we are told that they are the slanders prepared to justify a preconceived purpose of spoliation. No witness, it seems, will be admitted unless it be the witness of a friend. Unless some enemy of the Reformation can be found to confess the crimes which made the Reformation necessary, the crimes themselves are to be regarded as unproved. This is a hard condition. We appeal to Wolsey. Wolsey commenced the suppression. Wolsey first made public the infamies which disgraced the Church; while, notwithstanding, he died the devoted servant of the Church. This evidence is surely admissible? But no : Wolsey, too, must be put out of court. Wolsey was a courtier and a time-server. Wolsey was a tyrant's minion. Wolsey was—in short, we know not what Wolsey was, or what he was not. Who can put confidence in a charlatan? Behind the bulwarks of such objections, the champion of the abbeys may well believe himself secure.

And yet, unreasonable though these demands may be, it happens, after all, that we are able partially to gratify them. It is strange that, of all extant accusations against any one of the abbeys, the heaviest is from a quarter which even Lingard[2] himself would scarcely call suspicious. No picture left us by Henry's visitors surpasses, even if it equals, a description of the condition of the Abbey of St Albans, in the last quarter of the fifteenth century, drawn by Morton, Henry the Seventh's minister, Cardinal Archbishop, Legate of the Apostolic See, in a letter addressed by him to the Abbot of St Albans himself. We must request our reader's special attention for the next two pages.

In the year 1489, Pope Innocent the Eighth—moved with the enormous stories which reached his ear of the corruption of the houses of religion in England—granted a commission to the Archbishop of Canterbury to make inquiries whether these stories were true, and to proceed to correct and reform as might seem good to him. The regular clergy were exempt from episcopal visitation, except under especial direction from Rome. The occasion had appeared so serious as to make extraordinary interference necessary.

[2] [John Lingard (1771-1851), Roman Catholic historian of England.]

On the receipt of the Papal commission, Cardinal Morton, among other letters, wrote the following letter :—

John, by Divine permission, Archbishop of Canterbury, Primate of all England, Legate of the Apostolic See, to William, Abbot of the Monastery of St Albans, greeting.

We have received certain letters under seal, the copies whereof we herewith send you, from our most holy Lord and Father in Christ, Innocent, by Divine Providence Pope, the eighth of that name. We therefore, John, the Archbishop, the visitor, reformer, inquisitor, and judge therein mentioned, in reverence for the Apostolic See, have taken upon ourselves the burden of enforcing the said commission; and have determined that we will proceed by, and according to, the full force, tenor, and effect of the same.

And it has come to our ears, being at once publicly notorious and brought before us upon the testimony of many witnesses worthy of credit, that you, the abbot afore-mentioned, have been a long time noted and diffamed, and do yet continue so noted, of simony, of usury, of dilapidation and waste of the goods, revenues, and possessions of the said monastery, and of certain other enormous crimes and excesses hereafter written. In the rule, custody, and administration of the goods, spiritual and temporal, of the said monastery you are so remiss, so negligent, so prodigal, that whereas the said monastery was of old times founded and endowed by the pious devotion of illustrious princes, of famous memory, heretofore kings of this land, the most noble progenitors of our most serene Lord and King that now is, in order that true religion might flourish there, that the name of the Most High, in whose honour and glory it was instituted, might be duly celebrated there;

And whereas, in days heretofore, the regular observance of the said rule was greatly regarded, and hospitality was diligently kept;

Nevertheless, for no little time, during which you have presided in the same monastery, you and certain of your fellow-monks and brethren (whose blood, it is feared, through your neglect, a severe Judge will require at your hand) have relaxed the measure and form of religious life; you have laid aside the pleasant yoke of contemplation, and all regular observances—hospitality, alms, and those other offices of piety which of old time were exercised and ministered therein have decreased, and by your faults, your carelessness, your neglect and deed, do daily decrease more and more, and cease to be regarded— the pious vows of the founders are defrauded of their just intent—the ancient rule of your order is deserted; and not a few of your fellow-monks and brethern, as we most deeply grieve to learn, giving themselves over to a reprobate mind, laying aside the fear of God, do lead only a life of lasciviousness—nay, as is horrible to relate, be not afraid to defile the holy places, even the very churches of God, by infamous intercourse with nuns, &c. &c.

You yourself, moreover, among other grave enormities and abominable crimes whereof you are guilty, and for which you are noted and diffamed, have, in the first place, admitted a certain married woman, named Elena Germyn, who has separated herself without just cause from her husband, and for some time past has lived in adultery with another man, to be a nun or sister in the house or Priory of Bray, lying, as you pretend, within your jurisdiction. You have next appointed the same woman to be prioress of the said house, notwithstanding that her said husband was living at the time, and is still alive. And, finally, Father Thomas Sudbury, one of your brother monks, publicly, notoriously, and without interference or punishment from you, has associated, and still associates, with this woman as an adulterer with his harlot.

Moreover, divers other of your brethern and fellow-monks have resorted, and do resort, continually to her and other women at the same place, as to a public brothel or receiving house, and have received no correction therefor.

Nor is Bray the only house into which you have introduced disorder. At the nunnery of Sapwell, which you also contend to be under your jurisdiction, you change

the prioresses and superiors again and again at your own
will and caprice. Here, as well as at Bray, you depose
those who are good and religious; you promote to the
highest dignities the worthless and the vicious. The
duties of the order are cast aside; virtue is neglected;
and by these means so much cost and extravagance has
been caused, that to provide means for your indulgence
you have introduced certain of your brethren to preside
in their houses under the name of guardians, when in fact
they are no guardians, but thieves and notorious villains;
and with their help you have caused and permitted the
goods of the same priories to be dispensed, or to speak
more truly to be dissipated, in the above-described corruptions and other enormous and accursed offences. Those
places once religious are rendered and reputed as it were
profane and impious; and by your own and your creatures'
conduct, are so impoverished as to be reduced to the verge
of ruin.

In like manner, also, you have dealt with certain other
cells of monks which you say are subject to you, even
within the monastery of the glorious proto-martyr Alban[3]
himself. You have dilapidated the common property; you
have made away with the jewels; the copses, the woods,
the underwood, almost all the oaks, and other forest
trees, to the value of eight thousand marks and more, you
have made to be cut down without distinction, and they
have by you been sold and alienated. The brethren of the
abbey, some of whom, as is reported, are given over to all
the evil things of the world, neglect the service of God
altogether. They live with harlots and mistresses publicly
and continuously, within the precincts of the monastery
and without. Some of them, who are covetous of honour
and promotion, and desirous therefore of pleasing your
cupidity, have stolen and made way with the chalices and
other jewels of the church. They have even sacrilegiously
extracted the precious stones from the very shrine of St
Alban; and you have not punished these men, but have

[3] [St Alban, beheaded by the Romans about the year 304, was
reputed to be the first British martyr.]

rather knowingly supported and maintained them. If any of your brethren be living justly and religiously, if any be wise and virtuous, these you straightway depress and hold in hatred. . . . You . . .

But we need not transcribe further this overwhelming document. It pursues its way through mire and filth to its most lame and impotent conclusion. After all this, the Abbot was not deposed; he was invited merely to reconsider his doings, and, if possible, amend them. Such was Church discipline, even under an extraordinary commission from Rome. But the most incorrigible Anglican will scarcely question the truth of a picture drawn by such a hand; and it must be added that this one unexceptionable indictment lends at once assured credibility to the reports which were presented fifty years later, on the general visitation. There is no longer room for the presumptive objection that charges so revolting could not be true. We see that in their worst form they could be true, and the evidence of Legh and Leghton, of Rice, and Bedyll, as it remains in their letters to Cromwell, must be shaken in detail, or else it must be accepted as correct. We cannot dream that Archbishop Morton was mistaken, or was misled by false information. St Albans was no obscure priory in a remote and thinly-peopled county. The Abbot of St Albans was a peer of the realm, taking precedence of bishops, living in the full glare of notoriety, within a few miles of London. The Archbishop had ample means of ascertaining the truth; and, we may be sure, had taken care to examine his ground before he left on record so tremendous an accusation. This story is true—as true as it is piteous. We will pause a moment over it before we pass from this, once more to ask our passionate Church friends whether still they will persist that the abbeys were no worse under the Tudors than they had been in their origin, under the Saxons, or under the first Norman and Plantagenet kings. We refuse to believe it. The abbeys which towered in the midst of the English towns, the houses clustered at their feet like subjects round some majestic queen, were images indeed of the civil supremacy which the Church of the Middle Ages had asserted for itself; but they were images also of an inner spiritual sublimity, which had won the homage of grateful and admiring nations. The heavenly graces had once descended upon the monastic orders, making them ministers of mercy, patterns of celestial life, breathing witnesses of the power of the Spirit in renewing and sanctifying the heart. And then it was that art and wealth and genius poured out their treasures to raise fitting tabernacles for the dwelling of so divine a soul. Alike in the village and the city, amongst the unadorned walls and lowly roofs which closed in the humble dwellings of the laity, the majestic houses of the Father of mankind and of his especial servants rose up in sovereign beauty. And ever at the sacred gates sat Mercy, pouring out relief from a never-failing store to the poor and the suffering; ever within the sacred aisles the voices of holy men were pealing heavenwards in intercession for the sins of mankind; and such blessed influences were thought to exhale around those mysterious precincts, that even the poor outcasts of society— the debtor, the felon, and the outlaw—gathered round the walls as the sick men sought the shadow of the apostles, and lay there sheltered from the avenging hand, till their sins were washed from off their souls. The abbeys of the middle ages floated through the storms of war and conquest, like the ark upon the waves of the flood, in the midst of violence remaining inviolate, through the awful reverence which surrounded them. The abbeys, as Henry's visitors found them, were as little like what they once had been, as the living man in the pride of his growth is like the corpse which the earth makes haste to hide for ever.

The official letters which reveal the condition into which the monastic establishments had degenerated, are chiefly in the Cotton Library,[4] and a large number of them have been published by the Camden Society. Besides these, however, there are in the Rolls House[5] many other documents which confirm and complete the statements of the writers of those letters. There is a part of what seems to have been a digest of the ' Black Book '—an epitome of iniquities, under the title of the ' Compendium Compertorum.' There are also reports

[4] [Now in the British Museum.]
[5] [Now the Public Record Office.]

from private persons, private entreaties for inquiry, depositions of monks in official examinations, and other similar papers, which, in many instances, are too offensive to be produced, and may rest in obscurity, unless contentious persons compel us to bring them forward. Some of these, however, throw curious light on the habits of the time, and on the collateral disorders which accompanied the more gross enormities. They show us, too, that although the dark tints predominate, the picture was not wholly black : that as just Lot was in the midst of Sodom, yet was unable by his single presence to save the guilty city from destruction, so in the latest era of monasticism there were types yet lingering of an older and fairer age, who, nevertheless, were not delivered, like the patriarch, but perished most of them with the institution to which they belonged. The hideous exposure is not untinted with fairer lines; and we see traits here and there of true devotion, mistaken but heroic.

Of these documents, two specimens shall be given in this place, one of either kind; and both, so far as we know, new to modern history. The first is so singular, that we print it as it is found—a genuine antique, fished up, in perfect preservation, out of the wreck of the old world.

About eight miles from Ludlow, in the county of Herefordshire, once stood the Abbey of Wigmore. There was Wigmore Castle, a stronghold of the Welsh Marches, now, we believe, a modern, well-conditioned mansion; and Wigmore Abbey, of which we do not hear that there are any remaining traces. Though now vanished, however, like so many of its kind, the house was three hundred years ago in vigorous existence; and when the stir commenced for an inquiry, the proceedings of the Abbot of this place gave occasion to a memorial which stands in the Rolls collection as follows :[6]—

Articles to be objected against John Smart, Abbot of the Monastery of Wigmore, in the county of Hereford, to be exhibited to the Right Honourable Lord Thomas Cromwell, the Lord Privy Seal and Viceregent to the King's Majesty.

1. The said abbot is to be accused of simony, as well

[6] Rolls House MS., *Miscellaneous Papers*, First Series. 356.

for taking money for advocation[7] and putations[8] of benefices, as for giving of orders, or more truly, selling them, and that to such persons which have been rejected elsewhere, and of little learning and light consideration.

2. The said abbot hath promoted to orders many scholars when all other bishops did refrain to give such orders on account of certain ordinances devised by the King's Majesty and his Council for the common weal of this realm. Then resorted to the said abbot scholars out of all parts, whom he would promote to orders by sixty at a time, and sometimes more, and otherwhiles less. And sometimes the said abbot would give orders by night within his chamber, and otherwise in the church early in the morning, and now and then at a chapel out of the abbey. So that there be many unlearned and light priests made by the said abbot, and in the diocese of Llandaff, and in the places aforenamed—a thousand, as it is esteemed, by the space of this seven years he hath made priests, and received not so little money of them as a thousand pounds for their orders.

3. Item, that the said abbot now of late, when he could not be suffered to give general orders, for the most part doth give orders by pretence of dispensation; and by that colour he promoteth them to orders by two and three, and takes much money of them, both for their orders and for to purchase their dispensations after the time he hath promoted them to their orders.

4. Item, the said abbot hath hurt and dismayed his tenants by putting them from their leases, and by enclosing their commons from them, and selling and utter wasting of the woods that were wont to relieve and succour them.

5. Item, the said abbot hath sold corradyes,[9] to the damage of the said monastery.

[7] [Presentation to benefices.]

[8] [Possibly ' supposititious presentation to benefices,' implying some deceit or fraud.]

[9] [Pensions from ecclesiastical funds.]

6. Item, the said abbot hath alienated and sold the jewels and plate of the monastery, to the value of five hundred marks, *to purchase of the Bishop of Rome his bulls to be a bishop, and to annex the said abbey to his bishopric, to that intent that he should not for his misdeeds be punished, or deprived from his said abbey.*

7. Item, that the said abbot, long after that other bishops had renounced the Bishop of Rome, and professed them to the King's Majesty, did use, but more verily usurped, the office of a bishop by virtue of his first bulls purchased from Rome, till now of late, as it will appear by the date of his confirmation, if he have any.

8. Item, that he the said abbot hath lived viciously, and kept to concubines divers and many women that is openly known.

9. Item, that the said abbot doth yet continue his vicious living, as it is known, openly.

10. Item, that the said abbot hath spent and wasted much of the goods of the said monastery upon the aforesaid women.

11. Item, that the said abbot is malicious and very wrathful, not regarding what he saith or doeth in his fury or anger.

12. Item, that one Richard Gyles bought of the abbot and convent of Wigmore a corradye, and a chamber for him and his wife for term of their lives; and when the said Richard Gyles was aged and was very weak, he disposed his goods, and made executors to execute his will. And when the said abbot now being —— perceived that the said Richard Gyles was rich, and had not bequested so much of his goods to him as he would have had, the said abbot then came to the chamber of the said Richard Gyles, and put out thence all his friends and kinsfolk that kept him in his sickness; and then the said abbot set his brother and other of his servants to keep the sick man; and the night next coming after the said Richard Gyles's coffer was broken, and thence taken all

that was in the same, to the value of forty marks; and long after the said abbot confessed, before the executors of the said Richard Gyles, that it was his deed.

13. Item, that the said abbot, after he had taken away the goods of the said Richard Gyles, used daily to reprove and check the said Richard Gyles, and inquire of him where was more of his coin and money; and at the last the said abbot thought he lived too long, and made the sick man, after much sorry keeping, to be taken from his feather-bed, and laid upon a cold mattress, and kept his friends from him to his death.

15. Item, that the said abbot consented to the death and murdering of one John Tichkill, that was slain at his procuring, at the said monastery, by Sir Richard Cubley, canon and chaplain to the said abbot; which canon is and ever hath been since that time chief of the said abbot's council; and is supported to carry crossbowes, and to go whither he lusteth at any time, to fishing and hunting in the king's forests, parks, and chases; but little or nothing serving the quire, as other brethren do, neither corrected of the abbot for any trespass he doth commit.

16. Item, that the said abbot hath been perjured oft, as is to be proved and is proved; and as it is supposed, did not make a true inventory of the goods, chattels, and jewels of his monastery to the King's Majesty and his Council.

17. Item, that the said abbot hath infringed all the king's injunctions which were given him by Doctor Cave to observe and keep; and when he was denounced *in pleno capitulo* to have broken the same, he would have put in prison the brother as did denounce him to have broken the same injunctions, save that he was let by the convent there.

18. Item, that the said abbot hath openly preached against the doctrine of Christ, saying he ought not to love his enemy, but as he loves the devil; and that he should love his enemy's soul, but not his body.

19. Item, that the said abbot hath taken but small regard to the good-living of his household.

20. Item, that the said abbot hath had and hath yet a special favour to misdoers and manquellers,[10] thieves, deceivers of their neighbours, and by them [is] most ruled and counselled.

21. Item, that the said abbot hath granted leases of farms and advocations first to one man, and took his fine, and also hath granted the same lease to another man for more money; and then would make to the last taker a lease or writing, with an antedate of the first lease, which hath bred great dissension among gentlemen—as Master Blunt and Master Moysey, and other takers of such leases—and that often.

22. Item, the said abbot having the contrepaynes of leases in his keeping, hath, for money, rased out the number of years mentioned in the said leases, and writ a fresh number in the former taker's lease, and in the contrepayne thereof, to the intent to defraud the taker or buyer of the residue of such leases, of whom he hath received the money.

23. Item, the said abbot hath not, according to the foundation of his monastery, admitted needy tenants into certain alms-houses belonging to the said monastery; but of them he hath taken large fines, and some of them he hath put away that would not give him fines : whither poor, aged, and impotent people were wont to be freely admitted, and [to] receive the founder's alms that of the old customs [were] limited to the same—which alms is also diminished by the said abbot.

24. Item, that the said abbot did not deliver the bulls of his bishopric, that he purchased from Rome, to our sovereign lord the king's council till long after the time he had delivered and exhibited the bulls of his monastery to them.

25. Item, that the said abbot hath detained and yet doth detain servants' wages; and often when the said

[10] [Slayers of men.]

servants hath asked their wages, the said abbot hath put them into the stocks, and beat them.

26. Item, the said abbot, in times past, hath had a great devotion to ride to Llangarvan, in Wales, upon Lammas-day, to receive pardon there; and on the even he would visit one Mary Hawle, an old acquaintance of his, at the Welsh Poole, and on the morrow ride to the foresaid Llangarvan, to be confessed and absolved, and the same night return to company with the said Mary Hawle, at the Welsh Poole aforesaid, and Kateryn, the said Mary Hawle her first daughter, whom the said abbot long hath kept to concubine, and had children by her, that he lately married at Ludlow. And [there be] others that have been taken out of his chamber and put in the stocks within the said abbey, and others that have complained upon him to the king's council of the Marches of Wales; and the woman that dashed out his teeth, that he would have had by violence, I will not name now, nor other men's wives, lest it would offend your good lordship to read or hear the same.

27. Item, the said abbot doth daily embezzle, sell, and convey the goods and chattels, and jewels of the said monastery, having no need so to do, for it is thought that he hath a thousand marks or two thousand lying by him that he hath gotten by selling of orders, and the jewels and plate of the monastery and corradyes; and it is to be feared that he will alienate all the rest, unless your good lordship speedily make redress and provision to let the same.

28. Item, the said abbot was accustomed yearly to preach at Leyntwarden on the Festival of the Nativity of the Virgin Mary, where and when the people were wont to offer to an image there, and to the same the said abbot in his sermons would exhort them and encourage them. But now the oblation be decayed, the abbot, espying the image then to have a cote of silver plate and gilt, hath taken away of his own authority the said image, and the plate turned to his own use, and left his preaching there, saying it is no manner of profit to any man, and

the plate that was about the said image was named to be worth forty pounds.

29. Item, the said abbot hath ever nourished enmity and discord among his brethren; and hath not encouraged them to learn the laws and the mystery of Christ. But he that least knew was most cherished by him; and he hath been highly displeased and [hath] disdained when his brothers would say that ' it is God's precept and doctrine that ye ought to prefer before your ceremonies and vain constitutions.' This saying was high disobedient, and should be grievously punished; when that lying, obloquy, flattery, ignorance, derision, contumely, discord, great swearing, drinking, hypocrisy, fraud, superstition, deceit, conspiracy to wrong their neighbour, and other of that kind, was had in special favour and regard. Laud and praise be to God that hath sent us the true knowledge. Honour and long prosperity to our sovereign lord and his noble council, that teaches to advance the same. Amen.

By John Lee, your faithful bedeman, and canon of the said monastery of Wigmore.

Postcript.—My good lord, there is in the said abbey a cross of fine gold and precious stones, whereof one diamond was esteemed by Doctor Booth, Bishop of Hereford, worth a hundred marks. In that cross is enclosed a piece of wood, named to be of the cross that Christ died upon, and to the same hath been offering. And when it should be brought down to the church from the treasury, it was brought down with lights, and like reverence as should have been done to Christ Himself. I fear lest the abbot upon Sunday next, when he may enter the treasury, will take away the said cross and break it, or turn it to his own use, with other precious jewels that be there.

All these articles afore written be true as to the substance and true meaning of them, though peradventure for haste and lack of counsel, some words be set amiss or out of their place. That I will be ready to prove forasmuch as lies in me, when it shall like your honourable lordship to direct your commission to men (or any

man) that will be indifferent and not corrupt to sit upon
the same, at the said abbey, where the witnesses and
proofs be most ready and the truth is best known, or at
any other place where it shall be thought most convenient
by your high discretion and authority.

The statutes of Provisors, commonly called Præmunire
statutes, which forbade all purchases of bulls from Rome
under penalty of outlawry, have been usually considered in
the highest degree oppressive; and more particularly the public
censure has fallen upon the last application of those statutes,
when, on Wolsey's fall, the whole body of the clergy were
laid under a præmunire, and only obtained pardon on payment
of a serious fine. Let no one regret that he has learnt to be
tolerant to Roman Catholics as the nineteenth century knows
them. But it is a spurious charity which, to remedy a modern
injustice, hastens to its opposite; and when philosophic
historians indulge in loose invective against the statesmen of
the Reformation, they show themselves unfit to be trusted
with the custody of our national annals. The Acts of Parliament speak plainly of the enormous abuses which had grown
up under these bulls. Yet even the emphatic language of the
statutes scarcely prepares us to find an abbot able to purchase
with jewels stolen from his own convent a faculty to confer
holy orders, though there is no evidence that he had been
consecrated bishop, and to make a thousand pounds by selling
the exercise of his privileges. This is the most flagrant case
which has fallen under the eyes of the present writer. Yet it
is but a choice specimen out of many. He was taught to
believe, like other modern students of history, that the papal
dispensations for immorality, of which we read in Foxe
and other Protestant writers, were calumnies, but he has been
forced against his will to perceive that the supposed calumnies
were but the plain truth; he has found among the records—
for one thing, a list of more than twenty clergy in one diocese
who had obtained licences to keep concubines.[11] After some
experience, he advises all persons who are anxious to understand the English Reformation to place implicit confidence in
the Statute Book. Every fresh record which is brought to light

[11] Tanner MS. 105, Bodleian Library, Oxford.

is a fresh evidence in its favour. In the fluctuations of the conflict there were parliaments, as there were princes, of opposing sentiments; and the measures were passed, amended, repealed, or censured, as Protestants and Catholics came alternately into power. But whatever were the differences of opinion, the facts on either side which are stated in an Act of Parliament may be uniformly trusted. Even in the attainders for treason and heresy we admire the truthfulness of the details of the indictments although we deplore the prejudice which at times could make a crime of virtue.

We pass on to the next picture. Equal justice, or some attempt at it, was promised, and we shall perhaps part from the friends of the monasteries on better terms than they believe. At least, we shall add to our own history and to the Catholic martyrology a story of genuine interest.

We have many accounts of the abbeys at the time of their actual dissolution. The resistance or acquiescence of superiors, the dismissals of the brethren, the sale of the property, the destruction of relics, &c., are all described. We know how the windows were taken out, how the glass appropriated, how the ' melter ' accompanied the visitors to run the lead upon the roofs, and the metal of the bells, into portable forms. We see the pensioned regulars filing out reluctantly, or exulting in their deliverance, discharged from their vows, furnished each with his ' secular apparel,' and his purse of money, to begin the world as he might. These scenes have long been partially known, and they were rarely attended with anything remarkable. At the time of the suppression, the discipline of several years had broken down opposition, and prepared the way for the catastrophe. The end came at last, but as an issue which had been long foreseen.

We have sought in vain, however, for a glimpse into the interior of the houses at the first intimation of what was coming—more especially when the great blow was struck which severed England from obedience to Rome, and asserted the independence of the Anglican Church. Then, virtually, the fate of the monasteries was decided. As soon as the supremacy was vested in the Crown, inquiry into their condition could no longer be escaped or delayed; and then, through the length and breadth of the country, there must have been rare dismay. The account of the London Carthusians is indeed known to us, because they chose to die rather than yield submission where their consciences forbade them; and their isolated heroism has served to distinguish their memories. The Pope, as head of the Universal Church, claimed the power of absolving subjects from their allegiance to their king. He deposed Henry. He called on foreign princes to enforce his sentence; and, on pain of excommunication, commanded the native English to rise in rebellion. The king, in self-defence, was compelled to require his subjects to disclaim all sympathy with these pretensions, and to recognize no higher authority, spiritual or secular, than himself within his own dominions. The regular clergy throughout the country were on the Pope's side, secretly or openly. The Charterhouse monks, however, alone of all the order, had the courage to declare their convictions, and to suffer for them. Of the rest, we only perceive that they at last submitted; and since there was no uncertainty as to their real feelings, we have been disposed to judge them hardly as cowards. Yet we who have never been tried, should perhaps be cautious in our censures. It is possible to hold an opinion quite honestly, and yet to hesitate about dying for it. We consider ourselves, at the present day, persuaded honestly of many things; yet which of them should we refuse to relinquish if the scaffold were the alternative—or at least seem to relinquish, under silent protest?

And yet, in the details of the struggle at the Charterhouse, we see the forms of mental trial which must have repeated themselves among all bodies of the clergy wherever there was seriousness of conviction. If the majority of the monks were vicious and sensual, there was still a large minority labouring to be true to their vows; and when one entire convent was capable of sustained resistance, there must have been many where there was only just too little virtue for the emergency—where the conflict between interest and conscience was equally genuine, though it ended the other way. Scenes of bitter misery there must have been—of passionate emotion wrestling ineffectually with the iron resolution of the Government : and the faults of the Catholic party weigh so heavily against them in the course and progress of the Reformation, that we cannot willingly lose the few countervailing tints which soften the darkness of their conditions.

Nevertheless, for any authentic account of the abbeys at this crisis, we have hitherto been left to our imagination. A stern and busy administration had little leisure to preserve records of sentimental struggles which led to nothing. The Catholics did not care to keep alive the recollection of a conflict in which, even though with difficulty, the Church was defeated. A rare accident only could have brought down to us any fragment of a transaction which no one had an interest in remembering. That such an accident has really occurred, we may consider unusually fortunate. The story in question concerns the abbey of Woburn, and is as follows :—

At Woburn, as in many other religious houses, there were representatives of both the factions which divided the country; perhaps we should say of three—the sincere Catholics, the Indifferentists, and the Protestants. These last, so long as Wolsey was in power, had been frightened into silence, and with difficulty had been able to save themselves from extreme penalties. No sooner, however, had Wolsey fallen, and the battle commenced with the Papacy, than the tables turned, the persecuted became persecutors—or at least threw off their disguise—and were strengthened with the support of the large class who cared only to keep on the winning side. The mysteries of the faith came to be disputed at the public tables; the refectories rang with polemics; the sacred silence of the dormitories was broken for the first time by lawless speculation. The orthodox might have appealed to the Government; heresy was still forbidden by law, and, if detected, was still punished by the stake. But the orthodox among the regular clergy adhered to the Pope as well as to the faith, and abhorred the sacrilege of the Parliament as deeply as the new opinions of the Reformers. Instead of calling in the help of the law, they muttered treason in secret; and the Reformers, confident in the necessities of the times, sent reports to London of their arguments and conversations. The authorities in the abbey were accused of disaffection; and a commission of inquiry was sent down towards the end of the spring of 1536, to investigate. The depositions taken on this occasion are still preserved; and with the help of them, we can leap over three centuries of time, and hear the last echoes of the old monastic life in Woburn Abbey dying away in discord.

Where party feeling was running so high, there were, of course, passionate arguments. The Act of Supremacy, the spread of Protestantism, the power of the Pope, the state of England—all were discussed; and the possibilities of the future, as each party painted it in the colours of his hopes. The brethren, we find, spoke their minds in plain language, sometimes condescending to a joke.

Brother Sherbourne deposes that the sub-prior, 'on Candlemas-day last past (February 2, 1536), asked him whether he longed not to be at Rome where all his bulls were?' Brother Sherbourne answered that 'his bulls had made so many calves, that he had burned them. Whereunto the sub-prior said he thought there were more calves now than there were then.'

Then there were long and furious quarrels about 'my Lord Privy Seal' (Cromwell)—who was to one party, the incarnation of Satan; to the other, the delivering angel.

Nor did matters mend when from the minister they passed to the master.

Dan John Croxton being in 'the shaving-house' one day with certain of the brethren having their tonsures looked to, and gossiping as men do on such occasions, one 'Friar Lawrence did say that the king was dead.' Then said Croxton, 'Thanks be to God, his Grace is in good health, and I pray God so continue him;' and said further to the said Lawrence, 'I advise thee to leave thy babbling.' Croxton, it seems, had been among the suspected in earlier times. Lawrence said to him, 'Croxton, it maketh no matter what thou sayest, for thou art one of the new world;' whereupon hotter still the conversation proceeded. 'Thy babbling tongue,' Croxton said, 'will turn us all to displeasure at length.' 'Then,' quoth Lawrence, 'neither thou nor yet any of us all shall do well as long as we forsake our head of the Church, the Pope.' 'By the mass!' quoth Croxton, 'I hope thy Pope Roger were in thy belly, or thou in his, for thou art a false perjured knave to thy prince.' Whereunto the said Lawrence answered, saying, ' By the mass, thou liest! I was never sworn to forsake the Pope to be our head, and never will be.' ' Then,' quoth Croxton, ' thou shalt be sworn spite of thine heart one day, or I will know why nay.'

These and similar wranglings may be taken as specimens of the daily conversation at Woburn, and we can perceive how an abbot with the best intentions would have found it difficult to keep the peace. There are instances of superiors in other houses throwing down their command in the midst of the crisis in flat despair, protesting that their subject brethren were no longer governable. Abbots who were inclined to the Reformation could not manage the Catholics; Catholic abbots could not manage the Protestants; indifferent abbots could not manage either the one or the other. It would have been well for the Abbot of Woburn—or well as far as this world is concerned—if he, like one of these, had acknowledged his incapacity, and had fled from his charge.

His name was Robert Hobbes. Of his age and family, history is silent. We know only that he held his place when the storm rose against the Pope; that, like the rest of the clergy, he bent before the blast, taking the oath to the king, and submitting to the royal supremacy, but swearing under protest, as the phrase went, with the outward, and not with the inward man—in fact, perjuring himself. Though infirm, so far, however, he was too honest to be a successful counterfeit, and from the jealous eyes of the Neologians of the abbey he could not conceal his tendencies. We have significant evidence of the *espionage* which was established over all suspected quarters, in the conversations and trifling details of conduct on the part of the Abbot, which were reported to the Government.

In the summer of 1534, orders came that the Pope's name should be rased out wherever it was mentioned in the mass books. A malcontent, by name Robert Salford, deposed that ' he was singing mass before the Abbot at St Thomas's altar within the monastery, at which time he rased out with his knife the name of the canon.' The Abbot told him to ' take a pen and strike or cross him out.' The saucy monk said those were not the orders. They were to rase him out. ' Well, well,' the Abbot said, ' it will come again one day.' ' Come again, will it?' was the answer; ' if it do, then we will put him in again; but I trust I shall never see that day.' The mild Abbot could remonstrate, but could not any more command; and the proofs of his malignant inclinations were remembered against him for the ear of Cromwell.

In the general injunctions, too, he was directed to preach against the Pope, and to expose his usurpation; but he could not bring himself to obey. He shrank from the pulpit; he preached but twice after the visitation, and then on other subjects, while in the prayer before the sermon he refused, as we find, to use the prescribed form. He only said, ' You shall pray for the spirituality, the temporality, and the souls that be in the pains of purgatory; and did not name the king to be supreme head of the Church in neither of the said sermons, nor speak against the pretended authority of the Bishop of Rome.'

Again, when Paul the Third, shortly after his election, proposed to call a general council at Mantua, against which, by advice of Henry the Eighth, the Germans protested, we have a glimpse how eagerly anxious English eyes were watching for a turning tide. ' Hear you,' said the Abbot one day, ' of the Pope's holiness and the congregation of bishops, abbots, and princes gathered to the council at Mantua? They be gathered for the reformation of the Universal Church; and here now we have a book of the excuse of the Germans, by which we may know what heretics they be : for if they were Catholics and true men as they pretend to be, they would never have refused to come to a general council.'

So matters went with the Abbot for some months after he had sworn obedience to the king. Lulling his conscience with such opiates as the casuists could provide for him, he watched anxiously for a change, and laboured with but little reserve to hold his brethren to their old allegiance.

In the summer of 1535, however, a change came over the scene, very different from the outward reaction for which he was looking, and a better mind woke in the Abbot : he learnt that in swearing what he did not mean with reservations and nice distinctions, he had lied to heaven and lied to man :

that to save his miserable life he had perilled his soul. When the Oath of Supremacy was required of the nation, Sir Thomas More, Bishop Fisher, and the monks of the Charterhouse—mistaken, as we believe, in judgment, but true to their consciences, and disdaining evasion or subterfuge—chose, with deliberate nobleness, rather to die than to perjure themselves. This is no place to enter on the great question of the justice or necessity of those executions; but the story of the so-called martyrdoms convulsed the Catholic world. The Pope shook upon his throne; the shuttle of diplomatic intrigue stood still; diplomatists who had lived so long in lies that the whole life of man seemed but a stage pageant, a thing of show and tinsel, stood aghast at the revelation of English sincerity, and a shudder of great awe ran through Europe. The fury of party leaves little room for generous emotion, and no pity was felt for these men by the English Protestants. The Protestants knew well that if these same sufferers could have had their way, they would themselves have been sacrificed by hecatombs; and as they had never experienced mercy, so they were in turn without mercy. But to the English Catholics, who believed as Fisher believed, but who had not dared to suffer as Fisher suffered, his death and the death of the rest acted as a glimpse of the Judgment Day. Their safety became their shame and terror; and in the radiant example before them of true faithfulness, they saw their own falsehood and their own disgrace. So it was with Father Forest, who had taught his penitents in confession that they might perjure themselves, and who now sought a cruel death in voluntary expiation; so it was with Whiting, the Abbot of Glastonbury; so with others whose names should be more familiar to us than they are; and here in Woburn we are to see the feeble but genuine penitence of Abbot Hobbes. He was still unequal to immediate martyrdom, but he did what he knew might drag his death upon him if disclosed to the Government, and surrounded by spies he could have had no hope of concealment.

' At the time,' deposed Robert Salford, ' that the monks of the Charterhouse, with other traitors, did suffer death, the Abbot did call us into the Chapter-house, and said these words :—" Brethren, this is a perilous time; such a scourge was never heard since Christ's passion. Ye hear how good men suffer the death. Brethren, this is undoubted for our offences. Ye read, so long as the children of Israel kept the commandments of God, so long their enemies had no power over them, but God took vengeance of their enemies. But when they broke God's commandments, then they were subdued by their enemies, and so be we. Therefore let us be sorry for our offences. Undoubtedly he will take vengeance of our enemies; I mean those heretics that causeth so many good men to suffer thus. Alas, it is a piteous case that so much Christian blood should be shed. Therefore, good brethren, for the reverence of God, every one of you devoutly pray, and say this Psalm, ' O God, the heathen are come into thine inheritance; thy holy temple have they defiled, and made Jerusalem a heap of stones. The dead bodies of thy servants have they given to be meat to the fowls of the air, and the flesh of thy saints unto the beasts of the field. Their blood have they shed like water on every side of Jerusalem, and there was no man to bury them. We are become an open scorn unto our enemies, a very scorn and derision unto them that are round about us. Oh, remember not our old sins, but have mercy upon us, and that soon, for we are come to great misery. Help us, O God of our salvation, for the glory of thy name. Oh, be merciful unto our sins for thy name's sake. Wherefore do the heathen say, Where is now their God?' Ye shall say this Psalm," repeated the Abbot, " every Friday, after the litany, prostrate, when ye lie upon the high altar, and undoubtedly God will cease this extreme scourge." And so,' continues Salford, significantly, ' the convent did say this aforesaid Psalm until there were certain that did murmur at the saying of it, and so it was left.'

The Abbot, it seems, either stood alone, or found but languid support: even his own familiar friends whom he trusted, those with whom he had walked in the house of God, had turned against him; the harsh air of the dawn of a new world choked him: what was there for him but to die? But his conscience still haunted him: while he lived he must fight on, and so, if possible, find pardon for his perjury. The blows in those years fell upon the Church thick and fast. In

February, 1536, the Bill passed for the dissolution of the smaller monasteries; and now we find the sub-prior with the whole fraternity united in hostility, and the Abbot without one friend remaining.

' He did again call us together,' says the next deposition, ' and lamentably mourning for the dissolving the said houses, he enjoined us to sing " Salvator mundi, salva nos omnes," every day after lauds; and we murmured at it, and were not content to sing it for such cause; and so we did omit it divers days, for which the Abbot came unto the Chapter, and did in manner rebuke us, and said we were bound to obey his commandment by our profession, and so did command us to sing it again with the versicle " Let God arise, and let His enemies be scattered. Let them also that hate Him flee before Him." Also he enjoined us at every mass that every priest did sing, to say the collect, " Oh God, who despisest not the sighing of a contrite heart." And he said if we did this with good and true devotion, God would so handle the matter, that it should be to the comfort of all England, and so show us mercy as he showed unto the children of Israel. And surely, brethren, there will come to us a good man that will rectify these monasteries again that be now supprest, because " God can of these stones raise up children to Abraham." '

' Of the stones,' perhaps, but less easily of the stony-hearted monks, who, with pitiless smiles, watched the Abbot's sorrow, which should soon bring him to his ruin.

Time passed on, and as the world grew worse, so the Abbot grew more lonely. Desolate and unsupported, he was still unable to make up his mind the course which he knew to be right; but he slowly strengthened himself for the trial, and as Lent came on the season brought with it a more special call to effort; he did not fail to recognize it. The conduct of the fraternity sorely disturbed him. They preached against all which he most loved and valued, in language purposely coarse; and the mild sweetness of the rebukes which he administered, showed plainly on which side lay, in the Abbey of Woburn, the larger portion of the spirit of Heaven. Now, when the passions of those times have died away, and we can look back with more indifferent eyes, how touching is the following scene. There was one Sir William, curate of Woburn chapel, whose tongue, it seems, was rough beyond the rest. The Abbot met him one way, and spoke to him. ' Sir William,' he said, ' I hear tell ye be a great railer. I marvel that ye rail so. I pray you teach my cure the Scripture of God, and that may be to edification. I pray you leave such railing. Ye call the Pope a bear and a bandog. Either he is a good man or an ill. *Domino suo stat aut cadit.* The office of a bishop is honourable. What edifying is this to rail? Let him alone.'

But they would not let him alone, nor would they let the Abbot alone. He grew ' somewhat acrased,' they said; vexed with feelings of which they had no experience. He fell sick, sorrow and the Lent discipline weighing upon him. The brethren went to see him in his room; one Brother Dan Woburn came among the rest, and asked him how he did; the Abbot answered, ' I would that I had died with the good men that died for holding with the Pope. My conscience, my conscience doth grudge me every day for it.' Life was fast losing its value for him. What was life to him or any man when bought with a sin against his soul? ' If the Abbot be disposed to die, for that matter,' Brother Croxton observed, ' he may die as soon as he will.'

All Lent he fasted and prayed, and his illness grew upon him; and at length in Passion week he thought all was over, and that he was going away. On Passion Sunday he called the brethren about him, and as they stood round his bed, with their cold, hard eyes, ' he exhorted them all to charity,' he implored them ' never to consent to go out of their monastery; and if it chanced them to be put from it, they should in no wise forsake their habit.' After these words, ' being in a great agony, he rose out of his bed, and cried out and said, " I would to God, it would please Him to take me out of this wretched world; and I would I had died with the good men that have suffered death heretofore, for they were quickly out of their pain." '[12] Then, half wandering, he began to mutter

[12] Meaning, as he afterwards said, More and Fisher and the Carthusians.

to himself aloud the thoughts which had been working in him in his struggles; and quoting St Bernard's words about the Pope, he exclaimed, ' Tu qui es primatu Abel, gubernatione Noah, auctoritate Moses, judicatu Samuel, potestate Petrus, unctione Christus. Aliæ ecclesiæ habent super se pastores. Tu pastor pastorum es.'

Let it be remembered that this is no sentimental fiction begotten out of the brain of some ingenious novelist, but the record of the true words and sufferings of a genuine child of Adam, labouring in a trial too hard for him.

He prayed to die, and in good time death was to come to him; but not, after all, in the sick bed, with his expiation but half completed. A year before, he had thrown down the cross when it was offered him. He was to take it again—the very cross which he had refused. He recovered. He was brought before the council; with what result, there are no means of knowing. To admit the Papal supremacy when officially questioned was high treason. Whether the Abbot was constant, and received some conditional pardon, or whether his heart again for the moment failed him—whichever he did, the records are silent. This only we ascertain of him : that he was not put to death under the Statute of Supremacy. But, two years later, when the official list was presented to the Parliament of those who had suffered for their share in ' the Pilgrimage of Grace,' among the rest we find the name of Robert Hobbes, late Abbot of Woburn. To this solitary fact we can add nothing. The rebellion was put down, and in the punishment of the offenders there was unusual leniency; not more than thirty persons were executed, although forty thousand had been in arms. Those only were selected who had been most signally implicated. But they were all leaders in the movement; the men of highest rank, and therefore greatest guilt. They died for what they believed their duty; and the king and council did their duty in enforcing the laws against armed insurgents. He for whose cause each supposed themselves to be contending, has long since judged between them; and both parties perhaps now see all things with clearer eyes than was permitted to them on earth.

We also see more distinctly. We will not refuse the Abbot Hobbes a brief record of his trial and passion. And although twelve generations of Russells—all loyal to the Protestant ascendancy—have swept Woburn clear of Catholic associations, they, too, in these later days, will not regret to see revived the authentic story of its last Abbot.

ENGLAND'S FORGOTTEN WORTHIES[1]

The Reformation, the Antipodes, the American Continent, the Planetary system, and the infinite deep of the Heavens, have now become common and familiar facts to us. Globes and orreries are the playthings of our school-days; we inhale the spirit of Protestantism with our earliest breath of consciousness. It is all but impossible to throw back our imagination into the time when, as new grand discoveries, they stirred every mind which they touched with awe and wonder at the revelation which God had sent down among mankind. Vast spiritual and material continents lay for the first time displayed, opening fields of thought and fields of enterprise of which none could conjecture the limit. Old routine was broken up. Men were thrown back on their own strength and their own power, unshackled to accomplish whatever they might dare. And although we do not speak of these discoveries as the cause of that enormous force of heart and intellect which accompanied them (for they were as much the effect as the cause, and one reacted on the other), yet at any rate they afforded scope and room for the play of powers which, without such scope, let them have been as transcendent as they would, must have passed away unproductive and blighted.

An earnest faith in the supernatural, an intensely real conviction of the divine and devilish forces by which the universe was guided and misguided, was the inheritance of the Elizabethan age from Catholic Christianity. The fiercest and most lawless men did then really and truly believe in the actual personal presence of God or the devil in every accident, or scene, or action. They brought to the contemplation of the new heaven and the new earth an imagination saturated with the spiritual convictions of the old era, which were not lost, but only infinitely expanded. The planets, whose vastness they now learnt to recognize, were, therefore, only the more powerful for evil or for good; the tides were the breathing of

[1] *Westminster Review,* 1852.

Demogorgon; and the idolatrous American tribes were real worshippers of the real devil, and were assisted with the full power of his evil army.

It is a form of thought which, however in a vague and general way we may continue to use its phraseology, has become, in its detailed application to life, utterly strange to us. We congratulate ourselves on the enlargement of our understanding when we read the decisions of grave law courts in cases of supposed witchcraft : we smile complacently over Raleigh's story of the island of the Amazons, and rejoice that we are not such as he—entangled in the cobwebs of effete and foolish superstition. Yet the true conclusion is less flattering to our vanity. That Raleigh and Bacon could believe what they believed, and could be what they were notwithstanding, is to us a proof that the injury which such mistakes can inflict is unspeakably insignificant : and arising, as those mistakes arose, from a never-failing sense of the real awfulness and mystery of the world and of the life of human souls upon it, they witness to the presence in such minds of a spirit, the loss of which not the most perfect acquaintance with every law by which the whole creation moves can compensate. We wonder at the grandeur, the moral majesty of some of Shakespeare's characters, so far beyond what the noblest among ourselves can imitate, and at first thought we attribute it to the genius of the poet, who has outstripped nature in her creations. But we are misunderstanding the power and the meaning of poetry in attributing creativeness to it in any such sense. Shakespeare created, but only as the spirit of nature created around him, working in him as it worked abroad in those among whom he lived. The men whom he draws were such men as he saw and knew; the words they utter were such as he heard in the ordinary conversations in which he joined. At the Mermaid with Raleigh and with Sidney, and at a thousand unnamed English firesides, he found the living originals for his Prince Hals, his Orlandos, his Antonios, his Portias, his Isabellas. The closer personal acquaintance which we can form with the English of the age of Elizabeth, the more we are satisfied that Shakespeare's great poetry is no more than the rhythmic echo of the life which it depicts.

It was, therefore, with no little interest that we heard of the formation of a society which was to employ itself, as we understood, in republishing in accessible form some, if not all, of the invaluable records compiled or composed by Richard Hakluyt. Books, like everything else, have their appointed death-day: the souls of them, unless they be found worthy of a second birth in a new body, perish with the paper in which they lived; and the early folio Hakluyts, not from their own want of merit, but from the neglect of them, were expiring of old age. The five-volume quarto edition,[2] published in 1811, so little people then cared for the exploits of their ancestors, consisted but of 270 copies. It was intended for no more than for curious antiquaries, or for the great libraries, where it could be consulted as a book of reference; and among a people, the greater part of whom had never heard Hakluyt's name, the editors are scarcely to be blamed if it never so much as occurred to them that general readers would care to have the book within their reach.

And yet those five volumes may be called the Prose Epic of the modern English nation. They contain the heroic tales of the exploits of the great men in whom the new era was inaugurated; not mythic, like the Iliads and the Eddas, but plain broad narratives of substantial facts, which rival legend in interest and grandeur. What the old epics were to the royally or nobly born, this modern epic is to the common people. We have no longer kings or princes for chief actors, to whom the heroism like the dominion of the world had in time past been confined. But, as it was in the days of the Apostles, when a few poor fishermen from an obscure lake in Palestine assumed, under the Divine mission, the spiritual authority over mankind, so, in the days of our own Elizabeth, the seamen from the banks of the Thames and the Avon, the Plym and the Dart, self-taught and self-directed, with no impulse but what was beating in their own royal hearts, went out across the unknown seas fighting, discovering, colonizing, and graved out the channels, paving them at last with their bones, through which the commerce and enterprise of England

[2] [Now superseded by the edition in 20 vols, published by Messrs. Maclehose of Glasgow, 1905-7.]

has flowed out over all the world. We can conceive nothing, not the songs of Homer himself, which would be read among us with more enthusiastic interest than these plain massive tales; and a people's edition of them in these days, when the writings of Ainsworth and Eugène Sue circulate in tens of thousands, would perhaps be the most blessed antidote which could be bestowed upon us. The heroes themselves were the men of the people—the Joneses, the Smiths, the Davises, the Drakes; and no courtly pen, with the one exception of Raleigh, lent its polish or its varnish to set them off. In most cases the captain himself, or his clerk or servant, or some unknown gentleman volunteer sat down and chronicled the voyage which he had shared; and thus inorganically arose a collection of writings which, with all their simplicity, are for nothing more striking than for the high moral beauty, warmed with natural feeling, which displays itself through all their pages. With us, the sailor is scarcely himself beyond his quarter-deck. If he is distinguished in his profession, he is professional merely, or if he is more than that, he owes it not to his work as a sailor, but to independent domestic culture. With them, their profession was the school of their nature, a high moral education which most brought out what was most nobly human in them; and the wonders of earth, and air, and sea, and sky, were a real intelligible language in which they heard Almighty God speaking to them.

That such hopes of what might be accomplished by the Hakluyt Society should in some measure be disappointed, is only what might naturally be anticipated of all very sanguine expectation. Cheap editions are expensive editions to the publisher; and historical societies, from a necessity which appears to encumber all corporate English action, rarely fail to do their work expensively and infelicitously. Yet, after all allowances and deductions, we cannot reconcile ourselves to the mortification of having found but one volume in the series to be even tolerably edited, and that one to be edited by a gentleman to whom England is but an adopted country— Sir Robert Schomburgk. Raleigh's ' Conquest of Guiana,' with Sir Robert's sketch of Raleigh's history and character, form in everything but its cost a very model of an excellent volume.

For the remaining editors,[3] we are obliged to say that they have exerted themselves successfully to paralyze whatever interest was reviving in Hakluyt, and to consign their own volumes to the same obscurity to which time and accident were consigning the earlier editions. Very little which was really noteworthy escaped the industry of Hakluyt himself, and we looked to find reprints of the most remarkable of the stories which were to be found in his collection. The editors began unfortunately with proposing to continue the work where he had left it, and to produce narratives hitherto unpublished of other voyages of inferior interest, or not of English origin. Better thoughts appear to have occurred to them in the course of the work; but their evil destiny overtook them before their thoughts could get themselves executed. We opened one volume with eagerness, bearing the title of ' Voyages to the North-west,' in hope of finding our old friends Davis and Frobisher. We found a vast unnecessary Editor's Preface : and instead of the voyages themselves, which with their picturesqueness and moral beauty shine among the fairest jewels in the diamond mine of Hakluyt, we encountered an analysis and digest of their results, which Milton was called in to justify in an inappropriate quotation. It is much as if they had undertaken to edit ' Bacon's Essays,' and had retailed what they conceived to be the substance of them in their own language; strangely failing to see that the real value of the actions or the thoughts of remarkable men does not lie in the material result which can be gathered from them, but in the heart and soul of the actors or speakers themselves. Consider what Homer's ' Odyssey ' would be, reduced into an analysis.

The editor of the ' Letters of Columbus ' apologizes for the rudeness of the old seaman's phraseology. Columbus, he tells us, was not so great a master of the pen as of the art of navigation. We are to make excuses for him. We are put on our guard, and warned not to be offended, before we are introduced to the sublime record of sufferings under which a man of the highest order was staggering towards the end of his earthly calamities; although the inarticulate fragments in which his thought breaks out from him, are strokes of natural

[3] This essay was written 15 years ago.

art by the side of which literary pathos is poor and mean-
ingless.

And even in the subjects which they select they are pursued
by the same curious fatality. Why is Drake to be best known,
or to be only known, in his last voyage? Why pass over the
success, and endeavour to immortalize the failure? When
Drake climbed the tree in Panamá, he saw both oceans, and
vowed that he would sail a ship in the Pacific; when he
crawled out upon the cliffs of Terra Del Fuego, and leaned his
head over the southernmost angle of the world; when he scored
a furrow round the globe with his keel, and received the
homage of the barbarians of the antipodes in the name of the
Virgin Queen, he was another man from what he had become
after twenty years of court life and intrigue, and Spanish
fighting and gold-hunting. There is a tragic solemnity in his
end, if we take it as the last act of his career; but it is his
life, not his death, which we desire—not what he failed to do,
but what he did.

But every bad has a worse below it, and more offensive
than all these is the editor of Hawkins's ' Voyage to the South
Sea.' The narrative is striking in itself; not one of the best,
but very good; and, as it is republished complete, we can
fortunately read it through, carefully shutting off Captain
Bethune's notes with one hand, and we shall then find in it
the same beauty which breathes in the tone of all the writings
of the period.

It is a record of misfortune, but of misfortune which did
no dishonour to him who sunk under it; and there is a
melancholy dignity in the style in which Hawkins tells his
story, which seems to say, that though he had been defeated,
and had never again an opportunity of winning back his lost
laurels, he respects himself still for the heart with which he
endured a shame which would have broken a smaller man. It
would have required no large exertion of editorial self-denial
to have abstained from marring the pages with puns of which
' Punch ' would be ashamed, and with the vulgar affectation
of patronage with which the sea captain of the nineteenth
century condescends to criticize and approve of his half-
barbarous precursor. And what excuse can we find for such

an offence as this which follows?—The war of freedom of the Araucan Indians is the most gallant episode in the history of the New World. The Spaniards themselves were not behind-hand in acknowledging the chivalry before which they quailed, and after many years of ineffectual efforts, they gave up a conflict which they never afterwards resumed; leaving the Araucans alone, of all the American races with which they came in contact, a liberty which they were unable to tear from them. It is a subject for an epic poem; and whatever admiration is due to the heroism of a brave people whom no inequality of strength could appal and no defeats could crush, these poor Indians have a right to demand of us. The story of the war was well known in Europe: Hawkins, in coasting the western shores of South America, fell in with them, and the finest passage in his book is the relation of one of the incidents of the war :—

An Indian captain was taken prisoner by the Spaniards, and for that he was of name, and known to have done his devoir against them, they cut off his hands, thereby intending to disenable him to fight any more against them. But he, returning home, desirous to revenge this injury, to maintain his liberty, with the reputation of his nation, and to help to banish the Spaniard, with his tongue intreated and incited them to persevere in their accustomed valour and reputation, abasing the enemy and advancing his nation; condemning their contraries of cowardliness, and confirming it by the cruelty used with him and other his companions in their mishaps; showing them his arms without hands, and naming his brethren whose half feet they had cut off, because they might be unable to sit on horseback; with force arguing that if they feared them not, they would not have used so great inhumanity—for fear produceth cruelty, the companion of cowardice. Thus encouraged he them to fight for their lives, limbs, and liberty, choosing rather to die an honourable death fighting, than to live in servitude as fruitless members of the commonwealth. Thus using the office of a sergeant-major, and having loaden his two stumps with bundles of arrows, he succoured them who, in the succeeding battle, had their store wasted; and changing himself from place to place, animated and encouraged his countrymen with such comfortable persuasions, as it is reported and credibly believed, that he did more good with his words and presence, without striking a stroke, than a great part of the army did with fighting to the utmost.

It is an action which may take its place by the side of the myth of Mucius Scævola, or the real exploit of that brother of the poet Æschylus, who, when the Persians were flying from Marathon, clung to a ship till both his hands were hewn away, and then seized it with his teeth, leaving his name as a portent even in the splendid calendar of Athenian heroes. Captain Bethune, without call or need, making his notes, merely, as he tells us, from the suggestions of his own mind as he revised the proof-sheets, informs us, at the bottom of the page, that ' it reminds him of the familiar lines—

> *For Widdrington I needs must wail,*
> *As one in doleful dumps;*
> *For when his legs were smitten off,*
> *He fought upon his stumps.'*

It must not avail him, that he has but quoted from the ballad of Chevy Chase. It is the most deformed stanza[4] of the modern deformed version which was composed in the eclipse of heart and taste, on the restoration of the Stuarts; and if such verses could then pass for serious poetry, they have ceased to sound in any ear as other than a burlesque; the associations which they arouse are only absurd, and they could only have continued to ring in his memory through their ludicrous doggrel.

When to these offences of the Society we add, that in the long laboured appendices and introductions, which fill up

---

[4] Here is the old stanza. Let whoever is disposed to think us too hard on Captain Bethune compare them:—

> *For Wetharrington my harte was woe,*
> *That even he slayne sholde be;*
> *For when both his leggis were hewen in to,*
> *He knyled and fought on his knee.*

Even Percy, who, on the whole, thinks well of the modern ballad, gives up this stanza as hopeless.

valuable space, which increase the expense of the edition, and into reading which many readers are, no doubt, betrayed, we have found nothing which assists the understanding of the stories which they are supposed to illustrate—when we have declared that we have found what is most uncommon passed without notice, and what is most trite and familiar encumbered with comment—we have unpacked our hearts of the bitterness which these volumes have aroused in us, and can now take our leave of them and go on with our more grateful subject.

Elizabeth, whose despotism was as peremptory as that of the Plantagenets, and whose ideas of the English constitution were limited in the highest degree, was, notwithstanding, more beloved by her subjects than any sovereign before or since. It was because, substantially, she was the people's sovereign; because it was given to her to conduct the outgrowth of the national life through its crisis of change, and the weight of her great mind and her great place were thrown on the people's side. She was able to paralyze the dying efforts with which, if a Stuart had been on the throne, the representatives of an effete system might have made the struggle a deadly one; and the history of England is not the history of France, because the resolution of one person held the Reformation firm till it had rooted itself in the heart of the nation, and could not be again overthrown. The Catholic faith was no longer able to furnish standing ground on which the English or any other nation could live a manly and godly life. Feudalism, as a social organization, was not any more a system under which their energies could have scope to move. Thenceforward, not the Catholic Church, but any man to whom God had given a heart to feel and a voice to speak, was to be the teacher to whom men were to listen; and great actions were not to remain the privilege of the families of the Norman nobles, but were to be laid within the reach of the poorest plebeian who had the stuff in him to perform them. Alone, of all the sovereigns in Europe, Elizabeth saw the change which had passed over the world. She saw it, and saw it in faith, and accepted it. The England of the Catholic Hierarchy and the Norman Baron, was to cast its shell and to become the England of free thought and commerce and manufacture, which was to plough the ocean with its navies, and sow its colonies over the globe; and the first appearance of these enormous forces and the light of the earliest achievements of the new era shines through the forty years of the reign of Elizabeth with a grandeur which, when once its history is written, will be seen to be among the most sublime phenomena which the earth as yet has witnessed. The work was not of her creation; the heart of the whole English nation was stirred to its depths; and Elizabeth's place was to recognize, to love, to foster, and to guide. The Government originated nothing; at such a time it was neither necessary nor desirable that it should do so; but wherever expensive enterprises were on foot which promised ultimate good, and doubtful immediate profit, we never fail to find among the lists of contributors the Queen's Majesty, Burghley, Leicester, Walsingham. Never chary of her presence, for Elizabeth could afford to condescend, when ships were fitting in the river for distant voyages, the Queen would go down in her barge and inspect. Frobisher, who was but a poor sailor adventurer, sees her wave her handkerchief to him from the Greenwich Palace windows, and he brings her home a narwhal's horn for a present. She honoured her people, and her people loved her; and the result was that, with no cost to the Government, she saw them scattering the fleets of the Spaniards, planting America with colonies, and exploring the most distant seas. Either for honour or for expectation of profit, or from the unconscious necessity by which a great people, like a great man, will do what is right, and must do it at the right time, whoever had the means to furnish a ship, and whoever had the talent to command one, laid their abilities together and went out to pioneer, and to conquer, and to take possession, in the name of the Queen of the Sea. There was no nation so remote but what someone or other was found ready to undertake an expedition there, in the hope of opening a trade; and let them go where they would, they were sure of Elizabeth's countenance. We find letters written by her, for the benefit of nameless adventurers, to every potentate of whom she had ever heard—to the Emperors of China, Japan, and India, the Grand Duke of Russia, the Grand Turk, the Persian ' Sofee,' and other unheard-of Asiatic and African princes; whatever was to be done in England, or by English-men, Elizabeth assisted when she could, and admired when she could not.

The springs of great actions are always difficult to analyze— impossible to analyze perfectly—possible to analyze only very proximately; and the force by which a man throws a good action out of himself is invisible and mystical, like that which brings out the blossom and the fruit upon the tree. The motives which we find men urging for their enterprises seem often insufficient to have prompted them to so large a daring. They did what they did from the great unrest in them which made them do it, and what it was may be best measured by the results in the present England and America.

Nevertheless there was enough in the state of the world, and in the position of England, to have furnished abundance of conscious motive, and to have stirred the drowsiest minister of routine.

Among material occasions for exertion, the population began to outgrow the employment, and there was a necessity for plantations to serve as an outlet. Men who, under happier circumstances, might have led decent lives, and done good service, were now driven by want to desperate courses—' witness,' as Richard Hakluyt says, ' twenty tall fellows hanged last Rochester assizes for small robberies;' and there is an admirable paper addressed to the Privy Council by Christopher Carlile, Walsingham's son-in-law, pointing out the possible openings to be made in or through such plantations for home produce and manufacture.

Far below all such prudential economies and mercantile ambitions, however, lay a chivalrous enthusiasm which in these dull days we can hardly, without an effort, realize. The life-and-death wrestle between the Reformation and the old religion had settled in the last quarter of the sixteenth century into a permanent struggle between England and Spain. France was disabled. All the help which Elizabeth could spare barely enabled the Netherlands to defend themselves. Pro-testantism, if it conquered, must conquer on another field; and by the circumstances of the time the championship of the

Reformed faith fell to the English sailors. The sword of Spain was forged in the goldmines of Peru; the legions of Alva were only to be disarmed by intercepting the gold ships on their passage; and, inspired by an enthusiasm like that which four centuries before had precipitated the chivalry of Europe upon the East, the same spirit which in its present degeneracy covers our bays and rivers with pleasure yachts, then fitted out armed privateers, to sweep the Atlantic, and plunder and destroy Spanish ships wherever they could meet them.

Thus, from a combination of causes, the whole force and energy of the age was directed towards the sea. The wide excitement, and the greatness of the interests at stake, raised even common men above themselves; and people who in ordinary times would have been no more than mere seamen, or mere money-making merchants, appear before us with a largeness and greatness of heart and mind in which their duties to God and their country are alike clearly and broadly seen and felt to be paramount to every other.

Ordinary English traders we find fighting Spanish war ships in behalf of the Protestant faith. The cruisers of the Spanish main were full of generous eagerness for the conversion of the savage nations to Christianity. And what is even more surprising, sites for colonization were examined and scrutinized by such men in a lofty statesmanlike spirit, and a ready insight was displayed by them into the indirect effects of a wisely-extended commerce on every highest human interest.

Again, in the conflict with the Spaniards, there was a further feeling, a feeling of genuine chivalry, which was spurring on the English, and one which must be well understood and well remembered, if men like Drake, and Hawkins, and Raleigh are to be tolerably understood. One of the English Reviews, a short time ago, was much amused with a story of Drake having excommunicated a petty officer as a punishment for some moral offence; the reviewer not being able to see in Drake, as a man, anything more than a highly brave and successful buccaneer, whose pretences to religion might rank with the devotion of an Italian bandit to the Madonna. And so Hawkins, and even Raleigh, are regarded by superficial persons, who see only such outward circumstances of their

history as correspond with their own impressions. The high
nature of these men, and the high objects which they pursued,
will only rise out and become visible to us as we can throw
ourselves back into their times and teach our hearts to feel
as they felt. We do not find in the language of the voyagers
themselves, or of those who lent them their help at home,
any of that weak watery talk of ' protection of aborigines,'
which, as soon as it is translated into fact, becomes the most
active policy for their destruction, soul and body. But the
stories of the dealings of the Spaniards with the conquered
Indians, which were widely known in England, seem to have
affected all classes of people, not with pious passive horror,
but with a genuine human indignation. A thousand anecdotes
in detail we find scattered up and down the pages of Hakluyt,
who, with a view to make them known, translated Peter
Martyr's letters; and each commonest sailor-boy who had
heard these stories from his childhood among the tales of his
father's fireside, had longed to be a man, that he might go out
and become the avenger of a gallant and suffering people.
A high mission, undertaken with a generous heart, seldom
fails to make those worthy of it to whom it is given; and it
was a point of honour, if of nothing more, among the English
sailors, to do no discredit by their conduct to the greatness
of their cause. The high courtesy, the chivalry of the Spanish
nobles, so conspicuous in their dealings with their European
rivals, either failed to touch them in their dealings with
uncultivated idolaters, or the high temper of the aristocracy
was unable to restrain or to influence the masses of the soldiers.
It would be as ungenerous as it would be untrue, to charge
upon their religion the grievous actions of men who called
themselves the armed missionaries of Catholicism, when the
Catholic priests and bishops were the loudest in the indignation
with which they denounced them. But we are obliged to charge
upon it that slow and subtle influence so inevitably exercised
by any religion which is divorced from life, and converted
into a thing of form, or creed, or ceremony, or system—
which could permit the same men to be extravagant in a
sincere devotion to the Queen of Heaven, whose entire
lower nature, unsubdued and unaffected, was given up to

thirst of gold, and plunder, and sensuality. If religion does not make men more humane than they would be without it, it makes them fatally less so; and it is to be feared that the spirit of the Pilgrim Fathers, which had oscillated to the other extreme, and had again crystallized into a formal antinomian fanaticism, reproduced the same fatal results as those in which the Spaniards had set them their unworthy precedent. But the Elizabethan navigators, full for the most part with the large kindness, wisdom, gentleness, and beauty, bear names untainted, as far as we know, with a single crime against the savages of America; and the name of England was as famous in the Indian seas as that of Spain was infamous. On the banks of the Oronoko there was remembered for a hundred years the noble captain who had come there from the great Queen beyond the seas; and Raleigh speaks the language of the heart of his country, when he urges the English statesmen to colonize Guiana, and exults in the glorious hope of driving the white marauder into the Pacific, and restoring the Incas to the throne of Peru.

> Who will not be persuaded (he says) that now at length the great Judge of the world hath heard the sighs, groans, and lamentations, hath seen the tears and blood of so many millons of innocent men, women, and children, afflicted, robbed, reviled, branded with hot irons, roasted, dismembered, mangled, stabbed, whipped, racked, scalded with hot oil, put to the strapado, ripped alive, beheaded in sport, drowned, dashed against the rocks, famished, devoured by mastiffs, burned, and by infinite cruelties consumed, and purposeth to scourge and plague that cursed nation, and to take the yoke of servitude from that distressed people, as free by nature as any Christian?

Poor Raleigh! if peace and comfort in this world were of much importance to him, it was in an ill day that he provoked the revenge of Spain. The strength of England was needed at the moment at its own door; the Armada came, and there was no means of executing such an enterprise. And afterwards the throne of Elizabeth was filled by a Stuart, and Guiana was to be no scene of glory for Raleigh; rather, as later historians are pleased to think, it was the grave of his reputation.

But the hope burned clear in him through all the weary years of unjust imprisonment; and when he was a grey-headed old man, the base son of a bad mother used it to betray him. The success of his last enterprise was made the condition under which he was to be pardoned for a crime which he had not committed; and its success depended, as he knew, on its being kept secret from the Spaniards. James required of Raleigh on his allegiance a detail of what he proposed, giving him at the same time his word as a king that the secret should be safe with him. The next day it was sweeping out of the port of London in the swiftest of Spanish ships, with private orders to the Governor of St Thomas to provoke a collision when Raleigh should arrive there, which should afterwards cost him his heart's blood.

We modern readers may run rapidly over the series of epithets under which Raleigh has catalogued the Indian sufferings, hoping that they are exaggerated, seeing that they are horrible, and closing our eyes against them with swiftest haste; but it was not so when every epithet suggested a hundred familiar facts; and some of these (not resting on English prejudice, but on sad Spanish evidence, which is too full of shame and sorrow to be suspected) shall be given in this place, however old a story it may be thought; because, as we said above, it is impossible to understand the actions of these men, unless we are familiar with the feelings of which their hearts were full.

The massacres under Cortez and Pizarro, terrible as they were, were not the occasion which stirred the deepest indignation. They had the excuse of what might be called, for want of a better word, necessity, and of the desperate position of small bands of men in the midst of enemies who might be counted by millions. And in De Soto, when he burnt his guides in Florida (it was his practice, when there was danger of treachery, that those who were left alive might take warning); or in Vasco Nunnez, praying to the Virgin on the mountains of Darien, and going down from off them into the valleys to hunt the Indian caciques, and fling them alive to his bloodhounds; there was, at least, with all this fierceness and cruelty, a desperate courage which we cannot refuse to admire, and which mingles with and corrects our horror. It is the refinement of the Spaniard's cruelty in the settled and conquered provinces, excused by no danger and provoked by no resistance, the details of which witness to the infernal coolness with which it was perpetrated; and the great bearing of the Indians themselves under an oppression which they despaired of resisting, raises the whole history to the rank of a worldwide tragedy, in which the nobler but weaker nature was crushed under a malignant force which was stronger and yet meaner than itself. Gold hunting and lust were the two passions for which the Spaniards cared; and the fate of the Indian women was only more dreadful than that of the men, who were ganged and chained to a labour in the mines which was only to cease with their lives, in a land where but a little before they had lived a free and contented people, more innocent of crime than perhaps any people upon earth. If we can conceive what our own feelings would be—if, in the ' development of the mammalia,' some baser but more powerful race than man were to appear upon this planet, and we and our wives and children at our own happy firesides were degraded from our freedom, and become to them what the lower animals are to us, we can perhaps realize the feelings of the enslaved nations of Hispaniola.

As a harsh justification to slavery, it is sometimes urged that men who do not deserve to be slaves will prefer death to the endurance of it; and that if they prize their liberty, it is always in their power to assert it in the old Roman fashion. Tried even by so hard a rule, the Indians vindicated their right; and, before the close of the sixteenth century, the entire group of the Western Islands in the hands of the Spaniards containing, when Columbus discovered them, many millions of inhabitants, were left literally desolate from suicide. Of the anecdotes of this terrible self-immolation, as they were then known in England, here are a few out of many.

The first is simple, and a specimen of the ordinary method. A Yucatan cacique, who was forced with his old subjects to labour in the mines, at last ' calling those miners into an house, to the number of ninety-five, he thus debateth with them :'—

'My worthy companions and friends, why desire we to live any longer under so cruel a servitude? Let us now go unto the perpetual seat of our ancestors, for we shall there have rest from these intolerable cares and grievances which we endure under the subjection of the unthankful. Go ye before, I will presently follow you.' Having so spoken, he held out whole handfuls of those leaves which take away life, prepared for the purpose, and giving every one part thereof, being kindled to suck up the fume; who obeyed his command, the king and his chief kinsmen reserving the last place for themselves.

We speak of the crime of suicide, but few persons will see a crime in this sad and stately leave-taking of a life which it was no longer possible to bear with unbroken hearts. We do not envy the Indian who, with Spaniards before him as an evidence of the fruits which their creed brought forth, deliberately exchanged for it the old religion of his country, which could sustain him in an action of such melancholy grandeur. But the Indians did not always reply to their oppressors with escaping passively beyond their hands. Here is a story with matter in it for as rich a tragedy as Œdipus or Agamemnon; and in its stern and tremendous features, more nearly resembling them than any which were conceived even by Shakespeare.

An officer named Orlando had taken the daughter of a Cuban cacique to be his mistress. She was with child by him, but, suspecting her of being engaged in some other intrigue, he had her fastened to two wooden spits, not intending to kill her, but to terrify her; and setting her before the fire, he ordered that she should be turned by the servants of the kitchen.

The maiden, stricken with fear through the cruelty thereof, and strange kind of torment, presently gave up the ghost. The cacique, her father, understanding the matter, took thirty of his men and went to the house of the captain, who was then absent, and slew his wife, whom he had married after that wicked act committed, and the women who were companions of the wife, and her servants every one. Then shutting the door of the house, and putting fire under it, he burnt himself and all his companions that assisted him, together with the captain's dead family and goods.

This is no fiction or poet's romance. It is a tale of wrath and revenge, which in sober dreadful truth enacted itself upon this earth, and remains among the eternal records of the doings of mankind upon it. As some relief to its most terrible features, we follow it with a story which has a touch in it of diabolical humour.

The slave-owners finding their slaves escaping thus unprosperously out of their grasp, set themselves to find a remedy for so desperate a disease, and were swift to avail themselves of any weakness, mental or bodily, through which to retain them in life. One of these proprietors being informed that a number of his people intended to kill themselves on a certain day, at a particular spot, and knowing by experience that they were too likely to do it, presented himself there at the time which had been fixed upon, and telling the Indians when they arrived that he knew their intention, and that it was vain for them to attempt to keep anything a secret from him, he ended with saying, that he had come there to kill himself with them; that as he had used them ill in this world, he might use them worse in the next; 'with which he did dissuade them presently from their purpose.' With what efficacy such believers in the imortality of the soul were likely to recommend either their faith or their God; rather, how terribly all the devotion and all the earnestness with which the poor priests who followed in the wake of the conquerors laboured to recommend it were shamed and paralyzed, they themselves too bitterly lament.

It was idle to send out governor after governor with orders to stay such practices. They had but to arrive on the scene to become infected with the same fever; or if any remnant of Castilian honour, or any faintest echoes of the faith which they professed, still flickered in a few of the best and noblest, they could but look on with folded hands in ineffectual mourning; they could do nothing without soldiers, and the soldiers were the worst offenders. Hispaniola became a desert; the gold was in the mines, and there were no slaves left remaining to extract it. One means which the Spaniards dared to employ to supply the vacancy, brought about an incident which in its piteous pathos exceeds any story we have ever heard. Crimes and criminals are swept away by time, nature finds an antidote for their poison, and they and their ill consequences alike are blotted out and perish. If we do not forgive the villain at least we cease to hate him, as it grows more clear to us that he injures none so deeply as himself, but the θηριώδης κακία, the enormous wickedness by which humanity itself has been outraged and disgraced, we cannot forgive; we cannot cease to hate that; the years roll away, but the tints of it remain on the pages of history, deep and horrible as the way on which they were entered there.

> When the Spaniards understood the simple opinion of the Yucatan islanders concerning the souls of their departed, which, after their sins purged in the cold northern mountains should pass into the south, to the intent that, leaving their own country of their own accord, they might suffer themselves to be brought to Hispaniola, they did persuade those poor wretches, that they came from those places where they should see their parents and children, and all their kindred and friends that were dead, and should enjoy all kinds of delights with the embracements and fruition of all beloved beings. And they, being infected and possessed with these crafty and subtle imaginations, singing and rejoicing left their country, and followed vain and idle hope. But when they saw that they were deceived, and neither met their parents nor any that they desired, but were compelled to undergo grievous sovereignty and command, and to endure cruel and extreme labour, they either slew themselves, or, choosing to famish, gave up their fair spirits, being persuaded by no reason or violence to take food. So these miserable Yucatans came to their end.

It was once more as it was in the days of the Apostles. The New World was first offered to the holders of the old traditions. They were the husbandmen first chosen for the new vineyard, and blood and desolation were the only fruits which they reared upon it. In their hands it was becoming a kingdom, not of God, but of the devil, and a sentence of blight went out against them and against their works. How fatally it has worked, let modern Spain and Spanish America bear witness. We need not follow further the history of their dealings with the Indians. For their colonies, a fatality appears to have followed all attempts at Catholic colonization. Like shoots from an old decaying tree which no skill and no care can rear, they were planted, and for a while they might seem to grow; but their life was never more than a lingering death, a failure, which to a thinking person would outweigh in the arguments against Catholicism whole libraries of faultless *catenas*, and a *consensus patrum* unbroken through fifteen centuries for the supremacy of St Peter.

There is no occasion to look for superstitious causes to explain the phenomenon. The Catholic faith had ceased to be the faith of the large mass of earnest thinking capable persons; and to those who can best do the work, all work in this world sooner or later is committed. America was the natural home for Protestants; persecuted at home, they sought a place where they might worship God in their own way, without danger of stake or gibbet, and the French Huguenots, as afterwards the English Puritans, early found their way there. The fate of a party of Coligny's people, who had gone out as settlers, shall be the last of these stories, illustrating, as it does in the highest degree, the wrath and fury which the passions on both sides were boiling. A certain John Ribault, with about 400 companions, had emigrated to Florida. They were quiet inoffensive people, and lived in peace there several years, cultivating the soil, building villages, and on the best possible terms with the natives. Spain was at the time at peace with France; we are, therefore, to suppose that it was in pursuance of the great crusade, in which they might feel secure of the secret, if not the confessed, sympathy of the Guises, that a powerful Spanish fleet bore down upon this settlement. The French made no resistance, and they were seized and flayed alive, and their bodies hung out upon the trees, with an inscription suspended over them, ' Not as Frenchmen, but as heretics.' At Paris all was sweetness and silence. The settlement was tranquilly surrendered to the same men who had made it the scene of their atrocity; and two years

later, 500 of the very Spaniards who had been most active in
the murder were living there in peaceable possession, in two
forts which their relation with the natives had obliged them
to build. It was well that there were other Frenchmen living,
of whose consciences the Court had not the keeping, and who
were able on emergencies to do what was right without
consulting it. A certain privateer, named Dominique de
Gourges, secretly armed and equipped a vessel at Rochelle, and
stealing across the Atlantic and in two days collecting a strong
party of Indians, he came down suddenly upon the forts, and,
taking them by storm, slew or afterwards hanged every man he
found there, leaving their bodies on the trees on which they
had hanged the Huguenots, and with their own inscription
reversed against them,—' Not as Spaniards, but as murderers.'
For which exploit, well deserving of all honest men's praise,
Dominique de Gourges had to fly his country for his life;
and, coming to England, was received with honourable
welcome by Elizabeth.

It was at such a time, and to take their part amidst such
scenes as these, that the English navigators appeared along
the shores of South America, as the armed soldiers of the
Reformation, and as the avengers of humanity. As their enterprise was grand and lofty, so for the most part was the manner
in which they bore themselves worthy of it. They were no
nation of saints, in the modern sentimental sense of that word;
they were prompt, stern men—more ready ever to strike an
enemy than to parley with him; and, private adventurers as
they all were, it was natural enough that private rapacity and
private badness should be found among them as among
other mortals. Every Englishman who had the means was at
liberty to fit out a ship or ships, and if he could produce
tolerable vouchers for himself, received at once a commission
from the Court. The battles of England were fought by her
children, at their own risk and cost, and they were at liberty to
repay themselves the expense of their expeditions by plunder-
ing at the cost of the national enemy. Thus, of course, in a
mixed world, there were found mixed marauding crews of
scoundrels, who played the game which a century later was
played with such effect by the pirates of the Tortugas. Negro

hunters too, there were, and a bad black slave trade—in which Elizabeth herself, driven hard for money, did not disdain to invest her capital—but on the whole, and in the war with the Spaniards, as in the war with the elements, the conduct and character of the English sailors, considering what they were and the work which they were sent to do, present us all through that age with such a picture of gallantry, disinterestedness, and high heroic energy, as has never been overmatched; the more remarkable, as it was the fruit of no drill or discipline, no tradition, no system, no organized training, but was the free native growth of a noble virgin soil.

Before starting on an expedition, it was usual for the crew and the officers to meet and arrange among themselves a series of articles of conduct, to which they bound themselves by a formal agreement, the entire body itself undertaking to see to their observance. It is quite possible that strong religious profession, and even sincere profession, might be accompanied as it was in the Spaniards, with everything most detestable. It is not sufficient of itself to prove that their actions would correspond with it, but it is one among a number of evidences; and coming as most of these men come before us, with hands clear of any blood but of fair and open enemies, their articles may pass at least as indications of what they were.

Here we have a few instances :—

Richard Hawkins's ship's company was, as he himself informs us, an unusually loose one. Nevertheless, we find them ' gathered together every morning and evening to serve God;' and a fire on board, which only Hawkins's presence of mind prevented from destroying ship and crew together, was made use of by the men as an occasion to banish swearing out of the ship.

> With a general consent of all our company, it was ordained that there should be a palmer[5] or ferula which should be in the keeping of him who was taken with an oath; and that he who had the palmer should give to every one that he took swearing, a palmada with it and the ferula; and whosoever at the time of evening or morning prayer was found to have the palmer, should

[5] [Obsolete word meaning a rod.]

have three blows given him by the captain or the master; and that he should still be bound to free himself by taking another, or else to run in danger of continuing the penalty, which being executed a few days, reformed the vice, so that in three days together was not one oath heard to be sworn.

The regulations for Luke Fox's voyage commenced thus:—

For as much as the good success and prosperity of every action doth consist in the due service and glorifying of God, knowing that not only our being and preservation, but the prosperity of all our actions and enterprises, do immediately depend on His Almighty goodness and mercy; it is provided—

First, that all the company, as well officers as others, shall duly repair every day twice at the call of the bell to hear public prayers to be read, such as are authorized by the Church, and that in a godly and devout manner, as good Christians ought.

Secondly, that no man shall swear by the name of God, or use any profane oath, or blaspheme His holy name.

To symptoms such as these, we cannot but assign a very different value when they are the spontaneous growth of common minds, unstimulated by sense of propriety or rules of the service, or other official influence lay or ecclesiastic, from what attaches to the somewhat similar ceremonials in which, among persons whose position is conspicuous, important enterprises are now and then inaugurated.

We have said as much as we intend to say of the treatment by the Spaniards of the Indian women. Sir Walter Raleigh is commonly represented by historians as rather defective, if he was remarkable at all, on the moral side of his character. Yet Raleigh can declare proudly, that all the time he was on the Oronoko, 'neither by force nor other means had any of his men intercourse with any woman there;' and the narrator of the incidents of Raleigh's last voyage acquaints his correspondent 'with some particulars touching the government of the fleet, which, although other men in their voyages doubtless in some measure observed, yet in all the great volumes which have been written touching voyages, there is no precedent of so godly severe and martial government, which not only in itself is laudable and worthy of imitation, but is also fit to be written and engraven on every man's soul that coveteth to do honour to his country.'

Once more, the modern theory of Drake is, as we said above, that he was a gentleman-like pirate on a large scale, who is indebted for the place which he fills in history to the indistinct ideas of right and wrong prevailing in the unenlightened age in which he lived, and who therefore demands all the toleration of our own enlarged humanity to allow him to remain there. Let us see how the following incident can be made to coincide with this hypothesis :—

A few days after clearing the Channel on his first great voyage, he fell in with a small Spanish ship, which he took for a prize. He committed the care of it to a certain Mr Doughtie, a person much trusted by, and personally very dear to him, and this second vessel was to follow him as a tender.

In dangerous expeditions into unknown seas, a second smaller ship was often indispensable to success; but many finely intended enterprises were ruined by the cowardice of the officers to whom such ships were entrusted; who shrank as danger thickened, and again and again took advantage of darkness or heavy weather to make sail for England and forsake their commander. Hawkins twice suffered in this way; so did Sir Humfrey Gilbert; and, although Drake's own kind feeling for his old friend has prevented him from leaving an exact account of his offence, we gather from the scattered hints which are let fall, that he, too, was meditating a similar piece of treason. However, it may or may not have been thus. But when at Port St Julien, ' our General,' says one of the crew,—

> Began to inquire diligently of the actions of Mr Thomas Doughtie, and found them not to be such as he looked for, but tending rather to contention or mutiny, or some other disorder, whereby, without redresse, the success of the voyage might greatly have been hazarded. Whereupon the company was called together and made acquainted with the particulars of the cause, which were found, partly by Mr Doughtie's own confession, and

partly by the evidence of the fact, to be true, which, when our General saw, although his private affection to Mr Doughtie (as he then, in the presence of us all, sacredly protested) was great, yet the care which he had of the state of the voyage, of the expectation of Her Majesty, and of the honour of his country, did more touch him, as indeed it ought, than the private respect of one man; so that the cause being thoroughly heard, and all things done in good order as near as might be to the course of our law in England, it was concluded that Mr Doughtie should receive punishment according to the quality of his offence. And he, seeing no remedy but patience for himself, desired before his death to receive the communion, which he did at the hands of Mr Fletcher, our minister, and our General himself accompanied him in that holy action, which, being done, and the place of execution made ready, he, having embraced our General, and taken leave of all the company, with prayers for the Queen's Majesty and our realm, in quiet sort laid his head on the block, where he ended his life. This being done, our General made divers speeches to the whole company, persuading us to unity, obedience, love, and regard of our voyage, and for the better confirmation thereof, willed every man the next Sunday following to prepare himself to receive the communion, as Christian brethren and friends ought to do, which was done in very reverent sort, and so with good contentment every man went about his business.

The simple majesty of this anecdote can gain nothing from any comment which we might offer upon it. The crew of a common English ship organizing, of their own free motion, on that wild shore, a judgment hall more grand and awful than any most elaborate law court, is not to be reconciled with the pirate theory. Drake, it is true, appropriated and brought home a million and a half of Spanish treasure, while England and Spain were at peace. He took that treasure because for many years the officers of the Inquisition had made free at their pleasure with the lives and goods of English merchants and seamen. The king of Spain, when appealed to, had replied that he had no power over the Holy House; and it was necessary to make the king of Spain, or the Inquisition, or whoever were the parties responsible, feel that they could not play their pious pranks with impunity. When Drake seized the bullion at Panama, he sent word to the Viceroy that he should now learn to respect the properties of English subjects; and he added, that if four English sailors, who were prisoners in Mexico, were molested, he would execute 2,000 Spaniards and send the Viceroy their heads. Spain and England were at peace, but Popery and Protestantism were at war—deep, deadly, and irreconcileable.

Wherever we find them, they are still the same. In the courts of Japan or of China; fighting Spaniards in the Pacific, or prisoners among the Algerines; founding colonies which by-and-by were to grow into enormous Transatlantic republics, or exploring in crazy pinnaces the fierce latitudes of the Polar seas,—they are the same indomitable God-fearing men whose life was one great liturgy. ' The ice was strong, but God was stronger,' says one of Frobisher's men, after grinding a night and a day among the icebergs, not waiting for God to come down and split the ice for them, but toiling through the long hours, himself and the rest fending off the vessel with poles and planks, with death glaring at them out of the rocks. Icebergs were strong, Spaniards were strong, and storms, and corsairs, and rocks and reefs, which no chart had them noted—they were all strong; but God was stronger, and that was all which they cared to know.

Out of the vast number of illustrations it is difficult to make wise selections, but the attention floats loosely over generalities, and only individual instances can seize it and hold it fast. We shall attempt to bring our readers face to face with some of these men; not, of course, to write their biographies, but to sketch the details of a few scenes, in the hope that they may tempt those under whose eyes they may fall to look for themselves to complete the perfect figure.

Some two miles above the port of Dartmouth, once among the most important harbours in England, on a projecting angle of land which runs out into the river at the head of one of its most beautiful reaches, there has stood for some centuries the Manor House of Greenaway. The water runs deep all the way to it from the sea, and the largest vessels may ride with safety within a stone's throw of the windows. In the latter half of the sixteenth century there must have met, in the hall of this mansion, a party as remarkable as could have been found anywhere in England. Humfrey and Adrian Gilbert, with their half-brother, Walter Raleigh, here, when little boys, played at sailors in the reaches of Long Stream; in the summer evenings doubtless rowing down with the tide to the port, and wondering at the quaint figure-heads and carved prows of the ships which thronged it; or climbing on board, and listening, with hearts beating, to the mariners' tales of the new earth beyond the sunset. And here in later life, matured men, whose boyish dreams had become heroic action, they used again to meet in the intervals of quiet, and the rock is shown underneath the house where Raleigh smoked his first tobacco. Another remarkable man, of whom we shall presently speak more closely, could not fail to have made a fourth at these meetings. A sailor boy of Sandwich, the adjoining parish, John Davis, showed early a genius which could not have escaped the eye of such neighbours, and in the atmosphere of Greenaway he learned to be as noble as the Gilberts, and as tender and delicate as Raleigh. Of this party, for the present we confine ourselves to the host and owner, Humfrey Gilbert, knighted afterwards by Elizabeth. Led by the scenes of his childhood to the sea and to sea adventures, and afterwards, as his mind unfolded, to study his profession scientifically, we find him as soon as he was old enough to think for himself, or make others listen to him, ' amending the great errors of naval sea cards, whose common fault is to make the degree of longitude in every latitude of one common bigness;' inventing instruments for taking observations, studying the form of the earth, and convincing himself that there was a north-west passage, and studying the necessities of his country, and discovering the remedies for them in colonization and extended the markets for home manufactures. Gilbert was examined before the Queen's Majesty and the Privy Council, and the record of his examination he has himself left to us in a paper which he afterwards drew up, and strange enough reading it is. The most admirable conclusions stand side by side with the wildest conjectures.

Homer and Aristotle are pressed into service to prove that the ocean runs round the three old continents, and that America therefore is necessarily an island. The Gulf Stream, which he had carefully observed, eked out by a theory of the *primum mobile*, is made to demonstrate a channel to the north, corresponding to Magellan's Straits in the south, Gilbert believing, in common with almost every one of his day, that these straits were the only opening into the Pacific, and the land to the south was unbroken to the Pole. He prophesies a market in the East for our manufactured linen and calicoes :—

The Easterns greatly prizing the same, as appeareth in Hester, where the pomp is expressed of the great King of India, Ahasuerus, who matched the coloured clothes wherewith his houses and tents were apparelled, with gold and silver, as part of his greatest treasure.

These and other such arguments were the best analysis which Sir Humfrey had to offer of the spirit which he felt to be working in him. We may think what we please of them; but we can have but one thought of the great grand words with which the memorial concludes, and they alone would explain the love which Elizabeth bore him :—

Never, therefore, mislike with me for taking in hand any laudable and honest enterprise, for if through pleasure or idleness we purchase shame, the pleasure vanisheth, but the shame abideth for ever.

Give me leave, therefore, without offence, always to live and die in this mind : that he is not worthy to live at all that, for fear or danger of death, shunneth his country's service and his own honour, seeing that death is inevitable and the fame of virtue immortal, wherefore in this behalf *mutare vel timere sperno*.

Two voyages which he undertook at his own cost, which shattered his fortune, and failed, as they naturally might, since inefficient help or mutiny of subordinates, or other disorders, are inevitable conditions under which more or less great men must be content to see their great thoughts mutilated by the feebleness of their instruments, did not dishearten him, and in June 1583 a last fleet of five ships sailed from the port of Dartmouth, with commission from the Queen to discover and take possession from latitude 45° to 50° North—a voyage not a little noteworthy, there being planted in the course of it the first English colony west of the Atlantic. Elizabeth had a foreboding that she would never see him again. She sent him a jewel as a last token of her favour, and she desired Raleigh to have his picture taken before he went.

The history of the voyage was written by a Mr Edward Hayes, of Dartmouth, one of the principal actors in it, and as a composition it is more remarkable for fine writing than any very commendable thought in the author. But Sir Humfrey's nature shines through the infirmity of his chronicler; and in the end, indeed, Mr Hayes himself is subdued into a better mind. He had lost money by the voyage, and we will hope his higher nature was only under a temporary eclipse. The fleet consisted (it is well to observe the ships and the size of them) of the ' Delight,' 120 tons; the barque ' Raleigh,' 200 tons (this ship deserted off the Land's End); the ' Golden Hinde ' and the ' Swallow,' 40 tons each; and the ' Squirrel,' which was called the frigate, 10 tons. For the uninitiated in such matters, we may add, that if in a vessel the size of the last, a member of the Yacht Club would consider that he had earned a club-room immortality if he had ventured a run in the depth of summer from Cowes to the Channel Islands.

> We had in all (says Mr Hayes) 260 men, among whom
> we had of every faculty good choice. Besides, for solace
> to our own people, and allurement of the savages,
> we were provided of music in good variety, not omitting
> the least toys, as morris dancers, hobby horses, and
> May-like conceits to delight the savage people.

The expedition reached Newfoundland without accident. St John's was taken possession of, and a colony left there; and Sir Humfrey then set out exploring along the American coast to the south, he himself doing all the work in his little ten-ton cutter, the service being too dangerous for the larger vessels to venture on. One of these had remained at St John's. He was now accompanied only by the ' Delight ' and the ' Golden Hinde,' and these two keeping as near the shore as they dared. He spent what remained of the summer examining every creek and bay, marking the soundings, taking the bearings of the possible harbours, and risking his life, as every hour he was obliged to risk it in such a service, in thus leading, as it were, the forlorn hope in the conquest of the New World. How dangerous it was we shall presently see. It was towards the end of August.

The evening was fair and pleasant, yet not without token of storm to ensue, and most part of this Wednesday night, like the swan that singeth before her death, they in the 'Delight' continued in sounding of drums and trumpets and fifes, also winding the cornets and hautboys, and in the end of their jollity left with the battell and ringing of doleful knells.

Two days after came the storm; the 'Delight' struck upon a bank, and went down in sight of the other vessels, which were unable to render her any help. Sir Humfrey's papers, among other things, were all lost in her; at the time considered by him an irreparable misfortune. But it was little matter, he was never to need them. The 'Golden Hinde' and the 'Squirrel' were now left alone of the five ships. The provisions were running short, and the summer season was closing. Both crews were on short allowance; and with much difficulty Sir Humfrey was prevailed upon to be satisfied for the present with what he had done, and to lay off for England.

So upon Saturday, in the afternoon, the 31st of August, we changed our course, and returned back for England, at which very instant, even in winding about, there passed along between us and the land, which we now forsook, a very lion, to our seeming, in shape, hair, and colour; not swimming after the manner of a beast by moving of his feet, but rather sliding upon the water with his whole body, except his legs, in sight, neither yet diving under and again rising as the manner is of whales, porpoises, and other fish, but confidently showing himself without hiding, notwithstanding that we presented ourselves in open view and gesture to amaze him. Thus he passed along, turning his head to and fro, yawning and gaping wide, with ougly demonstration of long teeth and glaring eyes; and to bidde us farewell, coming right against the ' Hinde,' he sent forth a horrible voice, roaring and bellowing as doth a lion, which spectacle we all beheld so far as we were able to discern the same, as men prone to wonder at every strange thing. What opinion others had thereof, and chiefly the General himself, I forbear to deliver. But he took it for *Bonum Omen*, rejoicing that he was to war against such an enemy, if it were the devil.

We have no doubt that he did think it was the devil; men in those days believing really that evil was more than a principle or a necessary accident, and that in all their labour for God and for right, they must make their account to have to fight with the devil in his proper person. But if we are to call it superstition, and if this were no devil in the form of a roaring lion, but a mere great seal or sea-lion, it is a more innocent superstition to impersonate so real a power, and it requires a bolder heart to rise up against it and defy it in its living terror, than to sublimate it away into a philosophical principle, and to forget to battle with it in speculating on its origin and nature. But to follow the brave Sir Humfrey, whose work of fighting with the devil was now over, and who was passing to his reward. The 2nd of September the General came on board the ' Golden Hinde ' ' to make merry with us.' He greatly deplored the loss of his books and papers, but he was full of confidence from what he had seen, and talked with eagerness and warmth of the new expedition for the following year. Apocryphal gold-mines still occupying the minds of Mr Hayes and others, they were persuaded that Sir Humfrey was keeping to himself some such discovery which he had secretly made, and they tried hard to extract it from him. They could make nothing, however, of his odd, ironical answers, and their sorrow at the catastrophe which followed is sadly blended with disappointment that such a secret should have perished. Sir Humfrey doubtless saw America with other eyes than theirs, and gold-mines richer than California in its huge rivers and savannahs.

Leaving the issue of this good hope (about the gold),

(continues Mr Hayes), to God, who only knoweth the truth thereof, I will hasten to the end of this tragedy, which must be knit up in the person of our General, and as it was God's ordinance upon him, even so the vehement persuasion of his friends could nothing avail to divert him from his wilful resolution of going in his frigate; and when he was entreated by the captain, master, and others, his wellwishers in the 'Hinde' not to venture, this was his answer—'I will not forsake my little company going homewards, with whom I have passed so many storms and perils.'

Two thirds of the way home they met foul weather and terrible seas, 'breaking short and pyramid-wise.' Men who had all their lives 'occupied the sea' had never seen it more outrageous. 'We had also upon our mainyard an apparition of a little fier by night, which seamen do call Castor and Pollux.'

Monday the ninth of September, in the afternoon, the frigate was near cast away oppressed by waves, but at that time recovered, and giving forth signs of joy, the General, sitting abaft with a book in his hand, cried unto us in the 'Hinde' so often as we did approach within hearing, 'we are as near to heaven by sea as by land,' reiterating the same speech, well beseeming a soldier resolute in Jesus Christ, and I can testify that he was. The same Monday night, about twelve of the clock, or not long after, the frigate being ahead of us in the 'Golden Hinde,' suddenly her lights were out, whereof as it were in a moment we lost the sight, and withal our watch cried, 'The General was cast away,' which was too true.

Thus faithfully (concludes Mr Hayes, in some degree rising above himself) I have related this story, wherein some spark of the knight's virtues, though he be extinguished, may happily appear; he remaining resolute to a purpose honest and godly as was this, to discover, possess, and reduce unto the service of God and Christian piety, those remote and heathen countries of America. Such is the infinite bounty of God, who from every evil deriveth good, that fruit may grow in time of our travelling in these North-Western lands (as has it not grown?), and the crosses, turmoils, and afflictions, both in the preparation and execution of the voyage, did correct the intemperate humours which before we noted to be in this gentleman, and made unsavoury and less delightful his other manifold virtues.

Thus as he was refined and made nearer unto the image of God, so it pleased the Divine will to resume him unto Himself, whither both his and every other high and noble mind have always aspired.

Such was Sir Humfrey Gilbert; still in the prime of his years when the Atlantic swallowed him. Like the gleam of a landscape lit suddenly for a moment by the lightning, these few scenes flash down to us across the centuries : but what a life must that have been of which this was the conclusion! We have glimpses of him a few years earlier, when he won his spurs in Ireland—won them by deeds which to us seem terrible in their ruthlessness, but which won the applause of Sir Henry Sidney as too high for praise or even reward. Chequered like all of us with lines of light and darkness, he was, nevertheless, one of a race which has ceased to be. We look round for them, and we can hardly believe that the same blood is flowing in our veins. Brave we may still be, and strong perhaps as they, but the high moral grace which made bravery and strength so beautiful is departed from us for ever.

Our space is sadly limited for historical portrait painting; but we must find room for another of that Greenaway party whose nature was as fine as that of Gilbert, and who intellectually was more largely gifted. The latter was drowned in 1583. In 1585 John Davis left Dartmouth on his first voyage into the Polar seas; and twice subsequently he went again, venturing in small ill-equipped vessels of thirty or forty tons into the most dangerous seas. These voyages were as remarkable for their success as for the daring with which they were accomplished, and Davis's epitaph is written on the map of the world, where his name still remains to commemorate his discoveries. Brave as he was, he is distinguished by a peculiar and exquisite sweetness of nature, which, from many little facts of his life, seems to have affected every one with whom he came in contact in a remarkable degree. We find men, for the love of Master Davis, leaving their firesides to sail with him, without other hope or motion; we find silver bullets cast to shoot him in a mutiny; the hard rude natures of the mutineers being awed by something in his carriage which was not like that of a common man. He has written the account of one of his northern voyages himself; one of those, by-the-by, which the Hakluyt Society have mutilated; and there is an imaginative beauty in it, and a rich delicacy of expression, which is called out in him by the first sight of strange lands and things and people.

To show what he was, we should have preferred, if possible, to have taken the story of his expedition into the South Seas, in which, under circumstances of singular difficulty, he was deserted by Candish, under whom he had sailed; and after inconceivable trials from famine, mutiny, and storm, ultimately saved himself and his ship, and such of the crew as had chosen to submit to his orders. But it is a long history, and will not admit of being curtailed. As an instance of the stuff of which it was composed, he ran back in the black night in a gale of wind through the Straits of Magellan, *by a chart which he had made with the eye in passing up*. His anchors were lost or broken; the cables were parted. He could not bring up the ship; there was nothing for it but to run, and he carried her safe through along a channel often not three miles broad, sixty miles from end to end, and twisting like the reaches of a river.

For the present, however, we are forced to content ourselves with a few sketches out of the north-west voyages. Here is one, for instance, which shows how an Englishman could deal with the Indians. Davis had landed at Gilbert's Sound, and gone up the country exploring. On his return he found his crew loud in complaints of the thievish propensities of the natives, and urgent to have an example made of some of them. On the next occasion he fired a gun at them with blank cartridge; but their nature was still too strong for them.

Seeing iron (he says), they could in no case forbear

stealing; which, when I perceived, it did but minister to me occasion of laughter to see their simplicity, and I willed that they should not be hardly used, but that our company should be more diligent to keep their things, supposing it to be very hard in so short a time to make them know their evils.

In his own way, however, he took an opportunity of administering a lesson to them of a more wholesome kind than could be given with gunpowder and bullets. Like the rest of his countrymen, he believed the savage Indians in their idolatries to be worshippers of the devil. ' They are witches,' he says; ' they have images in great store, and use many kinds of enchantments.' And these enchantments they tried on one occasion to put in force against himself and his crew.

Being on shore on the 4th day of July, one of them made a long oration, and then kindled a fire, into which with many strange words and gestures he put divers things, which we supposed to be a sacrifice. Myself and certain of my company standing by, they desired us to go into the smoke. I desired them to go into the smoke, which they would by no means do. I then took one of them and thrust him into the smoke, and willed one of my company to tread out the fire, and spurn it into the sea, which was done to show them that we did contemn their sorceries.

It is a very English story—exactly what a modern Englishman would do; only, perhaps, not believing that there was any real devil in the case, which makes a difference. However, real or not real, after seeing him patiently put up with such an injury, we will hope the poor Greenlander had less respect for the devil than formerly.

Leaving Gilbert's Sound, Davis went on to the north-west, and in lat. 63° fell in with a barrier of ice, which he coasted for thirteen days without finding an opening. The very sight of an iceberg was new to all his crew; and the ropes and shrouds, though it was midsummer, becoming compassed with ice,—

The people began to fall sick and faint-hearted—whereupon, very orderly, with good discretion, they entreated me to regard the safety of mine own life, as well as the preservation of theirs; and that I should not, through overbouldness, leave their widows and fatherless children to give me bitter curses.

Whereupon, seeking counsel of God, it pleased His Divine Majesty to move my heart to prosecute that which I hope shall be to His glory, and to the contentation of every Christian mind.

He had two vessels—one of some burthen, the other a pinnace of thirty tons. The result of the counsel which he had sought was, that he made over his own large vessel to such as wished to return, and himself, ' thinking it better to die with honour than to return with infamy,' went on, with such volunteers as would follow him, in a poor leaky cutter, up the sea now in commemoration of that adventure called Davis's Straits. He ascended 4° North of the furthest known point, among storms and icebergs, when the long days and twilight nights alone saved him from being destroyed, and, coasting back along the American shore, he discovered Hudson's Straits, supposed then to be the long-desired entrance into the Pacific. This exploit drew the attention of Walsingham, and by him Davis was presented to Burleigh, ' who was also pleased to show him great encouragement.' If either these statesmen or Elizabeth had been twenty years younger, his name would have filled a larger space in history than a small corner of the map of the world; but if he was employed at all in the last years of the century, no *vates sacer* has been found to celebrate his work, and no clue is left to guide us. He disappears; a cloud falls over him. He is known to have commanded trading vessels in the Eastern seas, and to have returned five times from India. But the details are all lost, and accident has only parted the clouds for a moment to show us the mournful setting with which he, too, went down upon the sea.

In taking out Sir Edward Michellthorne to India, in 1604, he fell in with a crew of Japanese, whose ship had been burnt, drifting at sea, without provisions, in a leaky junk. He supposed them to be pirates, but he did not choose to leave them to so wretched a death, and took them on board;

and in a few hours, watching their opportunity, they murdered him.

As the fool dieth, so dieth the wise, and there is no difference; it was the chance of the sea, and the ill reward of a humane action—a melancholy end for such a man—like the end of a warrior, not dying Epaminondas-like on the field of victory, but cut off in some poor brawl or ambuscade. But so it was with all these men. They were cut off in the flower of their days, and few of them laid their bones in the sepulchres of their fathers. They knew the service which they had chosen, and they did not ask the wages for which they had not laboured. Life with them was no summer holiday, but a holy sacrifice offered up on duty, and what their Master sent was welcome. Beautiful is old age—beautiful as the slow-dropping mellow autumn of a rich glorious summer. In the old man, nature has fulfilled her work; she loads him with her blessings; she fills him with the fruits of a well-spent life; and, surrounded by his children and his children's children, she rocks him softly away to a grave, to which he is followed with blessings. God forbid we should not call it beautiful. It is beautiful, but not the most beautiful. There is another life, hard, rough, and thorny, trodden with bleeding feet and aching brow; the life of which the cross is the symbol; a battle which no peace follows, this side the grave; which the grave gapes to finish, before the victory is won; and—strange that it should be so, this is the highest life of man. Look back along the great names of history; there is none whose life has been other than this. They to whom it has been given to do the really highest work in this earth—whoever they are, Jew or Gentile, Pagan or Christian, warriors, legislators, philosophers, priests, poets, kings, slaves—one and all, their fate has been the same—the same bitter cup has been given them to drink. And so it was with the servants of England in the sixteenth century. Their life was a long battle, either with the elements or with men; and it was enough for them to fulfil their work, and to pass away in the hour when God had nothing more to bid them do. They did not complain, and why should we complain for them? Peaceful life was not what they desired, and an honourable death had no terrors for them. Theirs was the old Grecian spirit, and the great heart of the Theban poet[6] lived again in them :

Θανεῖν δ' οἷσιν ἀνάγκα, τί κέ τις ἀνώνυμον
γῆρας ἐν σκότῳ καθήμενος ἕψοι μάταν,
ἀπάντων καλῶν ἄμμορος; [7]

'Seeing,' in Gilbert's own brave words, 'that death is inevitable, and the fame of virtue is immortal; wherefore in this behalf *mutare vel timere sperno.*'

In the conclusion of these light sketches we pass into an element different from that in which we have been lately dwelling. The scenes in which Gilbert and Davis played out their high natures were of the kind which we call peaceful, and the enemies with which they contended were principally the ice and the wind, and the stormy seas and the dangers of unknown and savage lands. We shall close amidst the roar of cannon and the wrath and rage of battle. Hume, who alludes to the engagement which we are going to describe, speaks of it in a tone which shows that he looked at it as something portentous and prodigious; as a thing to wonder at—but scarcely as deserving the admiration which we pay to actions properly within the scope of humanity—and, as if the energy which was displayed in it was like the unnatural strength of madness. He does not say this, but he appears to feel it; and he scarcely would have felt it if he had cared more deeply to saturate himself with the temper of the age of which he was writing. All the time, all England and all the world rang with the story. It struck a deeper terror, though it was but the action of a single ship, into the hearts of the Spanish people; it dealt a more deadly blow upon their fame and moral strength than the destruction of the Armada itself; and in the direct results which arose from it, it was scarcely less disastrous to them. Hardly, as it seems to us, if the most glorious actions which are set like jewels in the history of mankind are weighed one against the other in the balance, hardly will those 300

---

[6] [Pindar.]

[7] [' Since it is inevitable that men should die, why should a man sit in darkness and vainly nurture an inglorious old age, with no share in any fair thing?' I am indebted for this translation to the kindness of Mr. E. C. Yorke, of New College, Oxford.]

Spartans who in the summer morning sat ' combing their long
hair for death ' in the passes of Thermopylæ, have earned a
more lofty estimate for themselves than this one crew of
modern Englishmen.

In August 1591, Lord Thomas Howard, with six English
line-of-battle ships, six victuallers, and two or three pinnaces,
was lying at anchor under the Island of Florez. Light in
ballast and short of water, with half his men disabled by
sickness, Howard was unable to pursue the aggressive purpose
on which he had been sent out. Several of the ships' crews
were on shore : the ships themselves ' all pestered and rom-
maging,' with everything out of order. In this condition they
were surprised by a Spanish fleet consisting of 53 men-of-war.
Eleven out of the twelve English ships obeyed the signal of
the admiral, to cut or weigh their anchors and escape as they
might. The twelfth, the ' Revenge,' was unable for the
moment to follow. Of her crew of 190, ninety were sick on
shore, and, from the position of the ship, there was some
delay and difficulty in getting them on board. The ' Revenge '
was commanded by Sir Richard Grenville, of Bideford, a man
well-known in the Spanish seas, and the terror of the Spanish
sailors; so fierce he was said to be, that mythic stories passed
from lip to lip about him, and, like Earl Talbot or Cœur de
Lion, the nurses at the Azores frightened children with the
sound of his name. ' He was of great revenues, of his own
inheritance,' they said, ' but of unquiet mind, and greatly
affected to wars;' and from his uncontrollable propensities for
blood-eating, he had volunteered his services to the Queen; ' of
so hard a complexion was he, that I (John Huighen von
Linschoten, who is our authority here, and who was with the
Spanish fleet after the action) have been told by divers
credible persons who stood and beheld him, that he would
carouse three or four glasses of wine, and take the glasses
between his teeth and crush them in pieces and swallow them
down.' Such Grenville was to the Spaniard. To the English
he was a goodly and gallant gentleman, who had never turned
his back upon an enemy, and was remarkable in that remarkable
time for his constancy and daring. In this surprise at
Florez he was in no haste to fly. He first saw all his sick

on board and stowed away on the ballast; and then, with no more than 100 men left him to fight and work the ship, he deliberately weighed, uncertain, as it seemed at first, what he intended to do. The Spanish fleet were by this time on his weather bow, and he was persuaded (we here take his cousin Raleigh's beautiful narrative, and follow it in Raleigh's words) ' to cut his mainsail and cast about, and trust to the sailing of the ship :'—

But Sir Richard utterly refused to turn from the enemy, alledging that he would rather choose to die than to dishonour himself, his country, and her Majesty's ship, persuading his company that he would pass through their two squadrons in spite of them, and enforce those of Seville to give him way : which he performed upon diverse of the foremost, who, as the mariners term it, sprang their luff, and fell under the lee of the ' Revenge.' But the other course had been the better; and might right well have been answered in so great an impossibility of prevailing : notwithstanding, out of the greatness of his mind, he could not be persuaded.

The wind was light; the ' San Philip,' ' a huge high-charged ship ' of 1500 tons, came up to windward of him, and, taking the wind out of his sails, ran aboard him.

After the ' Revenge ' was entangled with the ' San Philip,' four others boarded her, two on her larboard and two on her starboard. The fight thus beginning at three o'clock in the afternoon continued very terrible all that evening. But the great ' San Philip,' having received the lower tier of the ' Revenge,' shifted herself with all diligence from her sides, utterly misliking her first entertainment. The Spanish ships were filled with soldiers, in some 200, besides the mariners, in some 500, in others 800. In ours there were none at all, besides the mariners, but the servants of the commander and some few voluntary gentlemen only. After many enterchanged vollies of great ordinance and small shot, the Spaniards deliberated to enter the ' Revenge,' and made divers attempts, hoping to force her by the multitude of their armed soldiers and musketeers; but were still repulsed again and again, and at all times beaten back into their own ship or into the sea. In the beginning of the fight the 'George Noble,' of London, having received some shot through her by the Armadas, fell under the lee of the 'Revenge,' and asked Sir Richard what he would command him; but being one of the victuallers, and of small force, Sir Richard bade him save himself and leave him to his fortune.

This last was a little touch of gallantry, which we should be glad to remember with the honour due to the brave English sailor who commanded the 'George Noble;' but his name has passed away, and his action is an *in memoriam*, on which time has effaced the writing. All that August night the fight continued, the stars rolling over in their sad majesty, but unseen through the sulphurous clouds which hung over the scene. Ship after ship of the Spaniards came on upon the 'Revenge,' 'so that never less than two mighty galleons were at her side and aboard her,' washing up like waves upon a rock, and falling foiled and shattered back amidst the roar of the artillery. Before morning fifteen several Armadas had assailed her, and all in vain; some had been sunk at her side; and the rest, 'so ill approving of their entertainment, that at break of day they were far more willing to hearken to a composition, than hastily to make more assaults or entries.' 'But as the day increased,' says Raleigh, 'so our men decreased; and as the light grew more and more, by so much the more grew our discomfort, for none appeared in sight but enemies, save one small ship called the "Pilgrim," commanded by Jacob Whiddon, who hovered all night to see the success, but in the morning, bearing with the "Revenge," was hunted like a hare among many ravenous hounds—but escaped.'

All the powder in the 'Revenge' was now spent, all her pikes were broken, 40 out of her 100 men killed, and a great number of the rest wounded. Sir Richard, though badly hurt early in the battle, never forsook the deck till an hour before midnight; and was then shot through the body while his wounds were being dressed, and again in the head. His surgeon was killed while attending on him; the masts were lying over the side, the rigging cut or broken, the upper works all shot in pieces, and the ship herself, unable to move, was settling slowly in the sea, the vast fleet of Spaniards lying round her in a ring, like dogs round a dying lion, and wary of approaching him in his last agony. Sir Richard, seeing that it was past hope, having fought for fifteen hours, and ' having by estimation eight hundred shot of great artillery through him,' ' commanded the master gunner, whom he knew to be a most resolute man, to split and sink the ship, that thereby nothing might remain of glory or victory to the Spaniards; seeing in so many hours they were not able to take her, having had above fifteen hours' time, above ten thousand men, and fifty-three men-of-war to perform it withal; and persuaded the company, or as many as he could induce, to yield themselves unto God and to the mercy of none else; but as they had, like valiant resolute men, repulsed so many enemies, they should now shorten the honour of their nation by prolonging their own lives for a few hours or a few days.'

The gunner and a few others consented. But such δαιμονίη ἀρετή[8] was more than could be expected of ordinary seamen. They had dared do all which did become men, and they were not more than men. Two Spanish ships had gone down, above 1500 of their crew were killed, and the Spanish Admiral could not induce any one of the rest of his fleet to board the ' Revenge ' again, ' doubting lest Sir Richard would have blown up himself and them, knowing his dangerous disposition.' Sir Richard lying disabled below, the captain, ' finding the Spaniards as ready to entertain a composition as they could be to offer it,' gained over the majority of the surviving company; and the remainder then drawing back from the master gunner, they all, without further consulting their dying commander, surrendered on honourable terms. If unequal to the English in action, the Spaniards were at least as courteous in victory. It is due to them to say, that the conditions were faithfully observed; and ' the ship being marvellous unsavourie,' Alonzo de Baçon, the Spanish admiral, sent his boat to bring Sir Richard on board his own vessel.

Sir Richard, whose life was fast ebbing away, replied that

---

[8] [God-like virtue.]

'he might do with his body what he list, for that he esteemed
it not;' and as he was carried out of the ship he swooned, and
reviving again, desired the company to pray for him.

The admiral used him with all humanity, 'commending his
valour and worthiness, being unto them a rare spectacle, and
a resolution seldom approved.' The officers of the fleet,
too, John Higgins tells us, crowded round to look at him; and
a new fight had almost broken out between the Biscayans and
the 'Portugals,' each claiming the honour of having boarded
the 'Revenge.'

> In a few hours Sir Richard, feeling his end approaching,
> showed not any sign of faintness, but spake these words
> in Spanish, and said, 'Here die I, Richard Grenville,
> with a joyful and quiet mind, for that I have ended my
> life as a true soldier ought to do that hath fought for his
> country, queen, religion, and honour. Whereby my soul
> most joyfully departeth out of this body, and shall always
> leave behind it an everlasting fame of a valiant and true
> soldier that hath done his duty as he was bound to do.'
> When he had finished these or other such like words, he
> gave up the ghost with great and stout courage, and no
> man could perceive any sign of heaviness in him.

Such was the fight at Florez, in that August of 1591,
without its equal in such of the annals of mankind as the
thing which we call history has preserved to us; scarcely equal-
led by the most glorious fate which the imagination of Barrère
could invent for the 'Vengeur.' Nor did the matter end
without a sequel awful as itself. Sea battles have been often
followed by storms, and without a miracle; but with a miracle,
as the Spaniards and the English alike believed, or without
one, as we moderns would prefer believing, 'there ensued
on this action a tempest so terrible as was never seen or heard
the like before.' A fleet of merchantmen joined the Armada
immediately after the battle, forming in all 140 sail; and of
these 140, only 32 ever saw Spanish harbour. The rest
foundered, or were lost on the Azores. The men-of-war had
been so shattered by shot as to be unable to carry sail; and
the 'Revenge' herself, disdaining to survive her commander,
or as if to complete his own last baffled purpose, like Samson,

buried herself and her 200 prize crew under the rocks of St Michael's.

And it may well be thought and presumed (says John Huighen) that it was no other than a just plague purposely sent upon the Spaniards; and that it might be truly said, the taking of the 'Revenge' was justly revenged on them; and not by the might or force of man, but by the power of God. As some of them openly said in the Isle of Tèrceira, that they believed verily God would consume them, and that he took part with the Lutherans and heretics . . . . saying further, that so soon as they had thrown the dead body of the Vice-Admiral Sir Richard Grenville overboard, they verily thought that as he had a devilish faith and religion, and therefore the devil loved him, so he presently sunk into the bottom of the sea and down into hell, where he raised up all the devils to the revenge of his death, and that they brought so great a storm and torments upon the Spaniards, because they only maintained the Catholic and Romish religion. Such and the like blasphemies against God they ceased not openly to utter.

CHENEYS AND THE HOUSE OF RUSSELL[1]

'The gardener and his wife,' Mr Tennyson tells us, 'laugh at the claims of long descent.' If it be so, the laugh is natural, for our first parents were 'novi homines,' and could not appreciate what they did not possess. Nevertheless, in all nations which have achieved any kind of eminence, particular families have stood out conspicuously for generation after generation as representatives of political principles, as soldiers or statesmen, as ruling in their immediate neighbourhoods with delegated authority, and receiving homage voluntarily offered. They have furnished the finer tissues in the corporate body of the national life, and have given to society its unity and coherence. In times of war they have fallen freely on the battle-field. In times of discord and civil strife their most illustrious members have been the first to bleed on the scaffold. An English family, it has been said, takes rank according to the number of its members which have been executed. With men, as with animals and plants, peculiar properties are propagated by breeding. Each child who has inherited a noble name feels a special call to do no dishonour to it by unworthy actions. The family falls in pieces when its characteristics disappear. But, be the cause what it may, there is no instance, ancient or modern, of any long-protracted national existence where an order of aristocracy and gentry is not to be found preserving their identity, their influence, and their privileges of birth through century after century. They have no monopoly of genius. A gifted man rises out of the people, receiving his patent of nobility, as Burns said, 'direct from Almighty God.' He makes a name and a position for himself. But when the name is made, he hands it on, with distinction printed upon it, to his children and his children's children. More is expected from the sons of eminent parents than from other men, and if the transmitted quality is genuine more comes out of them. It is not talent. Talent is but

[1] Fraser's Magazine, 1879.

partially hereditary, if at all. The virtue that runs in the blood is superiority of courage or character; and courage and character far more than cleverness, are the conditions indispensable for national leaders. Thus without exception, in all great peoples, hereditary aristocracies have formed themselves, and when aristocracies have decayed or disappeared the state has degenerated along with them. The fall of a nobility may be a cause of degeneracy, or it may only be a symptom; but the phenomenon itself is a plain matter of fact, true hitherto under all forms of political constitution, monarchic, oligarchic, or republican. Republics have held together as long as they have been strung with patrician sinews; when the sinews crack the republic becomes a democracy, and the unity of the commonwealth is shivered into a heap of disconnected atoms, each following its own laws of gravitation towards its imagined interests. Athens and Rome, the Italian Republics, the great kingdoms which rose out of the wreck of the Roman Empire, tell the same story. The modern Spaniard reads the records of the old greatness of his country on the tombs of the Castilian nobles, and in the ruins of their palaces. They and the glory of the Spanish race have departed together. The Alvas and the Olivarez's, the Da Leyvas and Mendozas may have deserved their fall; but when they fell, and no others had arisen in their places, the nation fell also. Hitherto no great state has been able to sustain itself in a front place without aristocracy of some kind maintained on the hereditary principle. On this point the answer of history is uniform. The United States may inaugurate a new experience. With the one exception of the Adams's, the great men who have shown as yet in American history have no representatives to stand at present in the front political ranks. There are no Washingtons, no Franklins, no Jeffersons, no Clays or Randolphs, now governing States or leading debates in Congress. How long this will continue, how long the determination that all men shall start equal in the race of life will prevail against the instinctive tendences of successful men to perpetuate their names, is the most interesting of political problems. The American nationality is as yet too young for conclusions to be built on what it has done hitherto, or has forborne to do. We shall know better two centuries hence whether equality and the ballot-box provide better leaders for a people than the old methods of birth and training. France was cut in pieces in the revolution in 1793, and flung into the Medean caldron, expecting to emerge again with fresh vitality. The rash experiment has not succeeded up to this time, and here too we must wait for what her future will bring forth. So far the nations which have democratized themselves have been successful in producing indefinite quantities of money. If money and money-making will secure their stability, they may look forward hopefully—not otherwise.

We, too, have travelled far on the same road. We can continue to say, ' Thank God we have still a House of Lords,' but it is a House of Lords which is allowed to stand with a conditional tenure. It must follow, it must not lead, the popular will. It has been preserved rather as an honoured relic of a state of things which is passing away, than as representing any actual forces now existing. We should not dream of creating a hereditary branch of legislature if we had to begin over again; being there, let it remain as long as it is harmless. Nevertheless, great families have still a hold upon the country, either from custom or from a sense of their value. Fifty years are gone since the great democratic Reform Bill, yet the hereditary peers must still give their consent to every law which passes. Their sons and cousins form a majority in the House of Commons, and even philosophic Radicals doubt if the character of the House would be improved without men there whose position in society is secured, and who can therefore afford to be patriotic. How long a privileged order will hold its ground against the tendencies of the age depends upon itself, and upon the objects which it places before itself. If those who are within the lines retain, on the whole, a superior tone to those outside, and if access to the patrician order is limited to men who have earned admission there by real merit, the Upper House will be left in spite of ballot and universal suffrage, or perhaps by means of them, for generations to come. But the outlook is not without its ugly features, and should anything happen to stir the passions of the people as they were stirred half a century ago, the English peerage would scarcely live through another storm.

Whatever future may be in store for them, the past at any rate is their own, and they are honourably proud of it. The Roman preserved in his palace the ashes of his titled ancestors, and exhibited their images in his saloons. The English noble hangs the armour which was worn at Flodden or at Crecy in his ancestral hall. The trophies and relics of generations are among the treasures of his family. The stately portraits of his sires look down upon him from the walls of his dining-room. When he dies his desire is, like the prayer of the Hebrews, to be buried in the sepulchre of his fathers. There only is the fitting and peaceful close of a life honourably spent. There the first founder of the family and his descendants rest side by side, after time has ceased for each of them, to be remembered together by the curious who spell through their epitaphs, and to dissolve themselves into common dust. Occasionally, as a more emphatic memorial, the mausoleum becomes a mortuary chapel attached to some parish church or cathedral. The original purpose was of course that a priest, specially appointed, should say masses there immediately close to the spot where their remains were lying. The custom has outlived the purpose of it, and such chapels are to be met with in Protestant countries as often as in Catholic. The most interesting that I ever saw is that of the Mendozas in the cathedral at Burgos. It is the more affecting because the Mendozas have ceased to exist. Nothing survives of them save their tombs, which, splendid as they are, and of the richest materials, are characteristically free from meretricious ornament. There lie the figures of the proudest race in the whole nobility of Spain; knight and lady, prelate and cardinal. The stories of the lives of most of them are gone beyond recovery, and yet in those stone features can be read character as pure and grand as ever did honour to humanity. If a single family could produce so magnificent a group, we cease to wonder how Spain was once the sovereign of Europe, and the Spanish Court the home of courtesy and chivalry.

Next in interest to the monuments of the Mendozas, and second to them only because the Mendozas themselves are gone, are the tombs of the house of Russell in the chapel at Cheneys, in Buckinghamshire. The claims of the Russells to honourable memory the loudest Radical will acknowledge. For three centuries and a half they have led the way in what is called progress. They rose with the Reformation. They furnished a martyr for the Revolution of 1688. The Reform Bill is connected for ever with the name of Lord John. To know the biographies of the dead Russells is to know English history for twelve generations; and if the progress with which we are so delighted leads us safely into the Promised Land, as we are bound to believe that it will, Cheneys ought to become hereafter a place of pious pilgrimage.

The village stands on a chalk hill rising from the little river Ches, four miles from Rickmansworth, on the road to Amersham. The estate belongs to the Duke of Bedford, and is pervaded by an aspect of serene good manners, as if it was always Sunday. No vulgar noises disturb the general quiet. Cricket may be played there, and bowls and such games as propriety allows—but the oldest inhabitant can never have heard an oath spoken aloud, or seen a drunken man. Dirt and poverty are equally unknown. The houses, large and small, are solid and substantial, built of red brick, with high chimneys and pointed gables, and well-trimmed gardens before the doors. A Gothic fountain stands in the middle of the village green, under a cluster of tall elms, where picturesque neatly-dressed girls go for the purest water. Beyond the green a road runs, on one side of which stands the church and the parsonage, and on the other the remains of the once spacious manor house, which was built by the first Earl of Bedford on the site of an old castle of the Plantagenet kings. One wing of the manor house only survives, but so well constructed, and of material so admirable, that it looks as if it had been completed yesterday. In a field under the window is an oak which tradition says was planted by Queen Bess. More probably it is as old as the Conquest. The entire spot, church, mansion, cottages, and people, form a piece of ancient England artificially preserved from the intrusion of modern ways. No land is let on building lease in Cheneys to be disfigured by contractors' villas. No flaring shops, which such villas bring behind them, make the street hideous. A single miscellaneous store supplies the simple wants of the few inhabitants—the bars of soap, the bunches of dip candles, the tobacco in ounce packets, the tea, coffee, and sugar, the balls of twine, the strips of calico. Even the bull's-eyes and gingerbread for the children are not unpermitted, if they are honestly made and warranted not to be poisonous. So light is the business that the tidy woman who presides at the counter combines with it the duties of the post-office, which again are of the simplest kind. All is old-fashioned, grave, and respectable. No signs are to be found of competition, of the march of intellect, of emancipation, of the divine right of each man and woman to do what is good in their own eyes—of the blessed liberty which the House of Russell has been so busy in setting forward. The inhabitants of Cheneys live under authority. The voice of the Russells has been the voice of the emancipator—the hand has been the hand of the ruling noble.

The Manor House contains nothing of much interest. In itself, though a fragment, it is a fine specimen of the mason work of the Tudor times, and if not pulled down will be standing strong as ever when the new London squares are turned to dust heaps. With the high-pitched roofs and its clusters of curiously twisted chimneys it has served as a model for the architecture of the village, the smallest cottages looking as if they had grown from seeds which had been dropped by the central mansion.

All this is pretty enough, but the attraction of the place to a stranger is the church and what it contains. I had visited it before more than once, but I wished to inspect the monuments more closely. I ran down from London, one evening in June, to the village inn, and in the morning, soon after sunrise, when I was in less danger of having the officious assistance thrust upon me of clerk or sexton, I sauntered over to see if I could enter. The keys were kept at an adjoining cottage. The busy matron was already up and at her work. When I told her that I had special permission she unlocked the church door and left me to myself. Within, as without, all was order. No churchwardens, it was plain enough, had ever been allowed to work their will at Cheneys. Nay, the unchallenged loyalty of the Bedford family to constitutional liberty must have saved the Church from the visits of the Commissioners of the Long Parliament. On the walls are old Catholic brasses, one representing a parish priest of the place with the date of 1512, and a scroll praying for mercy on his soul. Strange to think that this man had said mass in the very place where I was standing, and that the memory of him had been preserved by the Russells, till the wheel had come round again and a Catholic hierarchy had been again established in England, with its Cardinals and Archbishops and Bishops. Will mass be ever said in Cheneys again?— not the sham mass of the Ritualists, but the real thing? Who that looks on England now can say that it will not? And four miles off is Amersham, where John Knox used to preach, and Queen Mary's inquisitors gathered their batches of heretics for Smithfield. On the pavement against the wall lies a stone figure of an old knight, finished only from the waist upwards. The knight is in his armour, his wife rests at his side; the hands of both of them reverently folded. Opening from the church on the north side, but private and not used for service, is the Russell Chapel. Below is the vault where the remains lie of most of the family who have borne the name for three centuries and a half.

On a stone tablet over the east window are the words ' This Chapel is built by Anne, Countess of Bedford, wife to John, Earl of Bedford, A.D. 1556.' It was the year in which Queen Mary was most busy offering her sacrifices to persuade Providence to grant her an heir. The chapel, therefore, by a curious irony, must have been consecrated with Catholic ceremonies.

The earliest monument is the tomb of this Lady Anne[2] and her husband, and is one of the finest of its kind in Europe. The material is alabaster; the pink veins in the stone being abundant enough to give a purple tint to the whole construction. The workmanship is extremely elaborate, and belongs to a time when the temper of men was still manly

[2] Through some blunder, she is described on the monument as Lady Elizabeth.

and stern, and when the mediæval reverence for death was still unspoiled by insincerity and affectation. The hands are folded in the old manner. The figures are not represented as sleeping, but as in a trance, with eyes wide open. The faces are evidently careful likenesses; the Earl has lost an eye in action—the lid droops over the socket as in life. His head rests on his corslet, his sword is at his side. He wears a light coronet and his beard falls low on his breast. The features do not denote a man of genius, but a loyal and worthy servant of the State, cautious, prudent, and thoughtful. The lady's face is more remarkable, and it would seem from the pains which have been taken with it that the artist must have personally known and admired her, while the Earl he may have known only by his portrait. The forehead of the Lady Anne is strong and broad, the nose large, the lips full but severely and expressively closed. She looks upward as she lies, with awe, but with a bold heart, stern as a Roman matron. The head is on a cushion, but the Earl's baldric would have formed as suitable a pillow for a figure so commanding and so powerful. It is a pity that we know so little of this lady. She was the daughter of Sir Guy Sapcote, of Huntingdonshire. Her mother was a Cheney, and through her the Cheneys estate fell to its present owners. She had been twice married and twice a widow when her hand was sought by Sir John Russell. At that time she was in the household of Catherine of Aragon; but she had no liking for the cause which Catherine represented, or for Catherine's daughter either. She died while Mary was still on the throne, but in her will she gave a significant proof that she at least had now bowed the knee when Baal was brought in again. She bequeathed her soul to Almighty God, ' trusting only by the death and passion of his dear Son, Jesus Christ, to be saved.' This is all that can be said of the ' mighty mother ' of the Russells to whose side they are gathered as they fall; but if the stern portrait speaks truth, her sons have inherited gifts from her more precious by far than the broad lands in Bedford and Huntingdon.

The Russells, or Rozels, are on the Battle Roll as having come from Normandy with the Conqueror. They played their part under the Plantagenets, not without distinction, and towards the end of the fifteenth century were a substantial family settled at Barwick, in Dorsetshire. In the year 1506, John, son and heir of the reigning head of the house, had returned from a tour on the continent bringing back with him accomplishments rare at all times with young proud Englishmen, and at that day unheard of save among the officially-trained clergy. Besides his other acquisitions he could speak French, and probably German. It happened that in that winter the Archduke Philip, with his mad wife Joanna, sister of Catherine of Arragon, was on his way from the Low Countries to Spain. As he was going down channel he was driven by a gale into Weymouth and having been extremely sea-sick, he landed to recover himself. Foreign princes are a critical species of guest. The relations of Henry VII with Joanna's father, Ferdinand, were just then on a doubtful footing. Prince Arthur was dead. Catherine was not yet married to his brother Henry, nor was it at all certain that she was to marry him; and when so great a person as the Archduke, and so nearly connected with Ferdinand, had come into England uninvited, the authorities in Dorsetshire feared to let him proceed on his voyage till their master's pleasure was known. A courier was despatched to London, and meanwhile Sir Thomas Trenchard, the most important gentleman in the neighbourhood, invited the whole party to stay with him at Wolverton Hall. Trenchard was Russell's cousin. His own linguistic capabilities were limited, and he sent for his young kinsman to assist in the royal visitors' entertainment. Russell went, and made himself extremely useful. Henry VII having pressed the Archduke to come to him at Windsor, the Archduke carried his new friend along with him, and spoke so warmly of his talents and character to the king that he was taken at once into the household. So commenced the new birth of the Russell house. Most men have chances opened to them at one time or another. Young Russell was one of the few who knew how to grasp opportunity by the forelock. He was found apt for any kind of service, either with pen or sword, brain or hand. He went with Henry VIII to his first campaign in

France. He was at the siege of Thérouenne, and at the battle of the Spurs. For an interval he was employed in political negotiations. Then we find him one of sixteen English knights who held the lists against all comers at Paris on the marriage of Louis XII with the Princess Mary. In the war of 1522 he lost his eye at the storming of Morlaix, and was knighted for his gallantry there. Immediately afterwards he was employed by Henry and Wolsey on an intricate and dangerous service. Louis XII was dead. The friendship between England and France was broken, and Henry and his nephew, the Emperor Charles V., were leagued together against the young Francis. Charles was aiming at the conquest of Italy. Henry had his eye on the French crown, which he dreamt of recovering for himself. Francis had affronted his powerful kinsman and subject, the Duke of Bourbon. Bourbon had intimated that if England would provide him with money to raise an army, he would recognize Henry as his liege lord, and John Russell was the person sent to ascertain whether Bourbon could be trusted to keep his word. Russell, it seems, was satisfied. The money was provided and was committed to Russell's care, and the great powers of Europe made their first plunge into the convulsions which were to last for more than a century. Little did Henry and Charles know what they were doing, or how often they would change partners before the game was over. Bourbon invaded Provence, Sir John Russell attending upon him with the English treasure. The war rolled across the Alps, and Russell saw the great battle fought at Pavia, where France lost all save honour, and the French king was the prisoner of the Emperor.

Then, if ever, was the time for Henry's dream to have been accomplished; but it became too clear that the throne of France was not at Bourbon's disposition, and that even if he had been willing and able to keep his word the Emperor had no intention of allowing him to keep it. Henry and Wolsey had both been foiled in the object nearest to their hearts, for Henry could not take the place of Francis, and Wolsey, who had meant to be pope, saw the Cardinal de Medici chosen instead of him. So followed a shift of policy. Charles V was now the danger to the rest of Europe. Henry joined himself with France against his late ally. Francis was to be liberated from his Spanish prison, and was to marry Henry's daughter. Catherine of Arragon was to be divorced, and Henry was to marry a French princess, or some one else in the French interest. The adroit Russell in Italy was to bring Milan, Venice, and the Papacy into the new confederacy. An ordinary politician looking then at the position of the pieces on the European chess-board, would have said that Charles, in spite of himself, would have been compelled to combine with the German princes, and to take up the cause of the Reformation. The Pope was at war with him. Clement, Henry, and Francis were heartily friends. Henry had broken a lance with Luther. Bourbon's army, which had conquered at Pavia, was recruited with lanzknechts, either Lutherans or godless ruffians. Bourbon's army was now Charles's; and food being scanty and pay not forthcoming, the Duke was driven, like another Alaric, to fling himself upon Rome, and storm and plunder the imperial city. It is curious and touching to find Clement clinging in such a hurricane to England and Henry as his surest supports. Russell had been staying with him at the Vatican on the eve of the catastrophe. He had gone home before the Germans approached, and missed being present at the most extraordinary scene in the drama of the sixteenth century, when the Holy Father, from the battlements of St. Angelo, saw his city sacked, his churches pillaged, his sacred sisterhoods outraged, his cardinals led in mockery on asses' backs through the streets by wild bands, acting under the order, or in the name, of the most Catholic King.

An attitude so extravagant could not endure. A little while, and the laws of spiritual attraction had forced the various parties into more appropriate relations. The divorce of Catherine went forward; the Pope fell back on Catherine's Imperial nephew. England broke with the Holy See, and the impulses which were to remodel the modern world flowed into their natural channels. Russell's friend, Thomas Cromwell, became Henry's chief minister; and Russell himself, though the scheme he had been employed to forward had burst like a bubble, still rose in his sovereign's confidence. He was at Calais with Henry in 1532 when Anne Boleyn was publicly received by Francis. He was active in the suppression of the monasteries, and presided at the execution of the Abbot of Glastonbury. Again, when Anne Boleyn fell into disgrace, Russell, who was now Privy Seal, was appointed with her uncle, the Duke of Norfolk, to examine into the charges against her. Through all the changes of Henry's later years, when the scaffold became so near a neighbour of the Royal closet, Russell remained always esteemed and trusted. At the birth of the young Edward he was made a peer, as Baron Russell of Cheneys. The year after he received the Garter. As Warden of the Stannaries he obtained the lands and mines of the suppressed Abbey of Tavistock. When his old master died he was carried on with the rising tide of the Reformation; he took Miles Coverdale for his chaplain, and obtained the Bishopric of Exeter for him. At his house in the Strand was held the conference on the Eucharist, when the strangest of all human superstitions was banished for a time from the English liturgy. Lord Russell's vigorous hand suppressed the Catholic rebellion in Devonshire. The Earldom of Bedford came next. His estates grew with his rank. Woburn Abbey fell to him on easy terms, for the Lords of the Council were first in the field, and had the pick of the spoil. Faction never tempted him out of the even road. He kept aloof from the quarrels of the Seymours and the Dudleys. When Somerset was attainted, the choicest morsel of Somerset's forfeited estates—Covent Garden and ' the seven acres '—was granted to the Earl of Bedford. Edward's death was a critical moment. Bedford, like the rest of his Council, signed the instrument for the succession of Lady Jane Grey. Like the rest, he changed his mind when he saw Lady Jane repudiated by the country. The blame of the conspiracy was thrown on the extreme Protestant faction. The moderate Liberals declared for Mary, and by retaining their places and their influence in the Council set limits to the reaction, and secured the next succession to Mary s sister. Mary's government became Catholic, but Russell continued Privy Seal. A rebellion broke out in Devonshire; this time a Protestant one. Bedford was the person who put it down. His last public act was to go with Lord Paget to

Spain to bring a Spanish husband home for his queen. He sailed with Philip from Corunna. He was at the memorable landing at Southampton, and he gave away his mistress at the marriage at Winchester. A few months later he died, after fifty years of service in the most eventful period of modern English history. His services were splendidly rewarded and he has been reproached in consequence as a trimmer and a time-server. But revolutions are only successful when they advance on a line lying between two extremes, and resulting from their compound action. To be a trimmer at such a time is to have discerned the true direction in which events are moving, and to be a wise man in whom good sense is stronger than enthusiasm. John Russell's lot was cast in an era of convulsion, when Europe was split into hostile camps, when religion was a shuttlecock of faction, Catholics and Protestants, as they were alternately uppermost, sending their antagonists to stake or scaffold. Russell represented the true feeling of the majority of Englishmen. They were ready to move with the age, to shake off the old tyranny of the Church, to put an end to monastic idleness, and to repudiate the authority of the Pope. But they had no inclination to substitute dogmatic Protestantism for dogmatic Catholicism. They felt instinctively that theologians knew but little, after all, of the subject for which they were so eager to persecute each other, and that the world had other interests beside those which were technically called religious; and on one point through all that trying time they were specially determined, that they would have no second war in England of rival Roses, no more fields of Towton or Barnet. They would work out their reformation, since a reformation there was to be, within the law and by the forms of it, and if enthusiasts chose to break into rebellion, or even passively to refuse obedience to the law like More or Fisher, they might be admired for their generous spirit, but they were struck down without hesitation or mercy. Who shall say that the resolution was not a wise one, or that men who acted upon it are proper objects of historical invective?

The mission to Spain rounds off John Russell's story. It commenced with his introduction to Philip's grandfather. It ended with Philip's marriage to the English Queen. Throughout his life his political sympathies were rather Imperial than French, as English feeling generally was. He was gone before the Marian persecution assumed its darker character; and until the stake became so busy, a wise liberal statesman might reasonably have looked on Mary's marriage with her cousin as promising peace for the country, and as a happy ending of an old quarrel.

Lady Anne lived to complete the Cheneys chapel; she died two years after her husband, and the Russells were then threatened with a change of fortune. The next Earl, Francis —Francis 'with the big head'—was born in 1528. His monument stands next that of his father and mother, and is altogether inferior to it. The two figures, the Earl himself and the Countess Margaret, are of alabaster like the first, and though wanting in dignity, are not in themselves wholly offensive; but according to the vile taste of the seventeenth century, they are tawdrily coloured in white and red and gold, and are lowered from the worthiness of sculpture to the level of a hair-dresser's model or of the painted Highlander at the door of a tobacco shop. Piety in England had by this time passed over to the Puritans, and Art, divorced from its proper inspiration, represented human beings as no better than wearers of State clothes. The Earl 'with the big head' deserves a more honourable portrait of himself, or deserves at least that the paint should be washed off. He was brought forward early in public life. He was Sheriff of Bedfordshire when he was nineteen. He sate in the Parliament of 1553, when the Prayer-book was purged of idolatry. In religion, taught perhaps by his mother, he was distinctly Protestant, and when his father died he was laid hold of as suspect by Gardiner. He escaped and joined the English exiles at Geneva. At the accession of Elizabeth he was called home, restored to his estates, and placed on the Privy Council, and when it pleased Mary Stuart, then Queen of France, to assume the royal arms of England, and declare herself the rightful owner of the English crown, the Earl of Bedford was sent to Paris to require that ambitious lady to limit those dangerous pretensions and to acknowledge her cousin's right.

Here it was that Bedford began his acquaintance with Mary Stuart; an acquaintance which was to be renewed under more agitating conditions. At Geneva, he had been intimate with the leading Reformers, Scotch as well as English. When Mary began her intrigues with the Catholic party in England, Bedford was sent to Berwick as Governor, where he could keep a watch over her doings, and be in constant communication with Knox and Murray. He received and protected Murray at the time of the Darnley marriage. Ruthven fled to him after the murder of Rizzio; and from Ruthven's lips Bedford wrote down the remarkable despatch, describing the details of the scene in that suite of rooms at Holyrood which has passed into our historical literature.

The Queen of Scots was regarded at this time by the great body of the English people as Elizabeth's indisputable heir. Catholic though she might be, her hereditary right was respected as Mary Tudor's had been, and had Elizabeth died while Darnley was alive, she would have succeeded as easily as James succeeded afterwards. When James was born he was greeted on his arrival in this world as a Prince of the Blood Royal, and Bedford was sent to Stirling to the christening with fine presents and compliments from his mistress. The shadow of the approaching tragedy hung over the ceremony. Bedford was conducted to the nursery to see the child in his cradle. Among the gifts which he had brought was a font of gold, which held the water in which James was made a Christian. Mary, in return, hung a chain of diamonds on Bedford's neck; never missing an opportunity of conciliating an English noble. But the English ambassador was startled to observe that the Queen's husband seemed of less consideration in her Court than the meanest footboy. The Queen herself scarce spoke to him; the courtiers passed him by with disdain. Bedford set it down to the murder of Rizzio, which he supposed to be still unforgiven, and he gave Mary a kindly hint that the poor wretch had friends in England whom in prudence she would do well to remember. Two months after came Kirk o' Field, and then the Bothwell marriage, Carberry Hill, Lochleven, Langside, the flight to England, the seventeen years in which the caged eagle beat her wings against her prison bars; and, finally, the closing scene in the hall at Fotheringay.

As his father had supported the rights of Mary Tudor, so the second Earl would have upheld the right of Mary Stuart till she had lost the respect of the country. But after Darnley's death the general sense of England pronounced her succession to be impossible. Bedford stood loyally by his own mistress in the dangers to which she was exposed from the rage of the disappointed Catholics. He was not one of the Lords of the Council who were chosen to examine the celebrated Casket letters, for he was absent at Berwick; but he sate on the trial of the Duke of Norfolk, and he joined in sending him to the scaffold. He died in 1585, two years before Mary Stuart's career was ended, but not before it was foreseen what that end must be. One other claim must not be forgotten which the second Earl possesses upon the memory of Englishmen. The famous Drake was born upon his estate at Tavistock. The Earl knew and respected his parents, and was godfather to their child, who derived from him the name of Francis. It was strange to feel that the actual remains of the man who had played a part in these great scenes were lying beneath the stones half a dozen yards from me. He sleeps sound, and the jangle of human discord troubles him no more.

He had two sons, neither of whom is in the vaults at Cheneys. Francis, the eldest, was killed while his father was alive, in a skirmish on the Scotch border. William fought at Zutphen by the side of Philip Sydney. For five years he was Viceroy of Ireland, which he ruled at least with better success than Essex, who came after him. This William was made Lord Russell of Thornhaugh, and brought a second peerage into the family. Their sister Anne was married to Ambrose Dudley, Earl of Warwick, the brother of Elizabeth's Leicester.

The third Earl, Edward, was the son of Francis who was killed in the north, and succeeded his grandfather when a boy of eleven. In him the family genius slept. He lived undistinguished and harmless, and died in 1627, having left unfulfilled even the simple duty of begetting an heir. He was followed by his cousin Francis, son of his uncle, Lord Thornhaugh, and the divided houses again became one.

This Francis was called the wise Earl. He was a true Russell, zealous for the Constitution and the constitutional liberties of England. He had been bred a lawyer, and understood all the arts of Parliamentary warfare. At the side of Eliot, and Pym, and Selden, he fought for the Petition of Right, and carried it by his own energy through the House of Lords. Naturally he made himself an object of animosity to the Court, and he was sent to the Tower as a reward of his courage. They could not keep him as they kept Eliot, to die there. He was released, but the battle had to be waged with weapons which a Russell was not disposed to use. When he was released Parliamentary life in England was suspended. There was no place for a Russell by the side of Laud and Strafford, and Bedford set himself to improve his property and drain the marshes about Whittlesea and Thorney. If solid work well done, if the addition of hundreds of thousands of acres to the soil available for the support of English life, be a title to honourable remembrance, this Earl ranks not the lowest in the Cheneys pantheon. He and his countess lie in the vault, with several of their children who died in childhood; they are commemorated in a monument not ungraceful in itself, were not it too daubed with paint and vulgarized by gilding. One of the little ones is a baby, a bambino swaddled round with wrappings which had probably helped to choke the infant life out of it.

The wise Earl died immediately after the opening of the Long Parliament. William Russell, his eldest son, had been returned to the House of Commons along with Pym as member for Tavistock. The Bedford interest doubtless gave Pym his seat there. His father's death removed him from the stormy atmosphere of the Lower House, and he was unequal to the responsibilities which his new position threw upon him. Civil war was not a theatre on which any Russell was likely to distinguish himself, and Earl William less than any of them. The old landmarks were submerged under the deluge. He was washed from side to side, fighting alternately in the field for King and Parliament. He signed the Covenant in

1645, but he found Woburn a pleasanter place than the council chamber, and thenceforward, till Cromwell's death, He looked on and took little part in public life. Charles twice visited him; once on his way back to Oxford after his failure at Chester, and again in 1647 when he was in the hands of the army, then quartered between Bedford and St Albans. It was at the time of the army manifesto, when the poor King imagined that he could play off Cromwell against the Parliament, and in fact was playing away his own life. After the negotiations were broken off, Charles went from Woburn to Latimers, a place close to Cheneys, from the windows of which, in the hot August days, he must have looked down on the Cheneys valley and seen the same meadows that now stretch along the bottom, and the same hanging beech woods, and the same river sparkling among its flags and rushes, and the cattle standing in the shallows. The world plunges on upon its way; generation follows generation, playing its part, and then ending. The quiet earth bears with them one after the other, and while all else changes, itself is changed so little.

This Earl was memorable rather from what befell him than from anything which he did. He was the first duke and he was the father of Lord William, whom English constitutional history has selected to honour as its chief saint and martyr. The Russells were not a family which was likely to furnish martyrs. They wanted neither courage, nor general decision of character, but they were cool and prudent; never changing their colours, but never rushing on forlorn hopes, or throwing their lives away on ill-considered enterprises.

Lord William, or Lord Russell, as he should be called, had perhaps inherited some exceptional quality in his blood. His mother was the beautiful Anne Carr, daughter of Carr, Earl of Somerset, the favourite of James I, and of Frances Howard, the divorced wife of the Earl of Essex, the hero and heroine of the great Oyer of poisoning, with its black surroundings of witchcraft and devilry. The old Earl Francis had sate upon their trial. He had been horrified when his son had proposed to marry the child of so ominous a pair. But Lady Anne was not touched by the crimes of her parents. Her loveliness shone perhaps the more attractively against so dark a background. Her character must have been singularly innocent, for she grew up in entire ignorance that her mother had been tried for murder. The family opposition was reluctantly withdrawn, and young Russell married her.

This pair, Earl William—afterwards Duke—and the Lady Anne Carr, are the chief figures in the most ostentatious monument in the Russell chapel. They are seated opposite each other in an attitude of violent grief, their bodies flung back, their heads buried in their hands in the anguish of petrified despair. They had many children, medallions of whom are ranged on either side in perpendicular rows. In the centre is the eldest—the occasion of the sorrow so conspicuously exhibited—whose head fell in Lincoln's Inn Fields. The execution of this medallion is extremely good; the likeness—if we may judge from the extant portraits of Lord Russell—is very remarkable. The expression is lofty and distinguished, more nearly resembling that of the first Countess than that of any of her other descendants; but there is a want of breadth, and the features are depressed and gloomy. It is a noble face, yet a face which tells of aspirations and convictions unaccompanied with the force which could carry them out into successful action. It stands with a sentence of doom upon it, the central object in a group of sculpture which, as a whole, is affected and hysterical. A man so sincere and so honourable deserves a simpler memorial, but it is not uncharacteristic of the pretentiousness and unreality which have been the drapery of the modern Whigs—their principles good and true in themselves, but made ridiculous by the extravagance of self-laudation.

Lord Russell's wife is a beautiful figure in the story, and she lies by his side in the Cheneys vault. She was Rachel Wriothesley, daughter of Lord Southampton; her mother being a De Rouvigny, one of the great Huguenot families in France. The tragedy of Lord Russell scarcely needs repeating. The Restoration was an experiment, to try whether the liberties of England were compatible with the maintenance of a dynasty which was Catholic at heart, and was for ever leaning as far as the times would permit to an avowal of Catholic belief.

Charles II had been obliged to hide his real creed, and pretend to Protestantism as a condition of his return. But the Catholic party grew daily stronger. Charles had no son, and the Duke of York was not Catholic only, but fanatically Catholic. Lord Russell led the opposition in Parliament. He shared to the bottom of his heart in the old English dread and hatred of Popery. He impeached Buckingham and Arlington. He believed to the last in the reality of the Popish plot, and he accepted Oates and Dangerfield as credible witnesses. He carried a Bill prohibiting Papists from sitting in Parliament. If Papists could not sit in Parliament, still less ought they to be on the throne, and the House of Commons, under his influence, passed the Exclusion Bill, cutting off the Duke of York. Russell carried it with his own hands to the House of Lords, and session after session, dissolution after dissolution, he tried to force the Lords to agree to it. No wonder that the Duke of York hated him, and would not spare him when he caught him tripping. When constitutional opposition failed, a true Russell would have been content to wait. But the husband of Lady Rachel drifted into something which, if not treason, was curiously like it, and under the shadow of his example a plot was formed by ruder spirits to save the nation by killing both the Duke and the King. Lord Russell was not privy to the Rye House affair, but he admitted that he had taken part in a consultation for putting the country in a condition to defend its liberties by force, and the enemy against whom the country was to be on its guard was the heir to the crown.

Martyrs may be among the best of men, but they are not commonly the wisest. To them their particular theories or opinions contain everything which makes life of importance, and no formula ever conceived by man is of such universally comprehensive character that it must be acted upon at all hazards and regardless of time and opportunity. The enthusiast imagines that he alone has the courage of his convictions; but there is a faith, and perhaps a deeper faith, which can stand still and wait till the fruit is ripe, when it can be gathered without violence. Each has its allotted part. The noble generous spirit sacrifices itself and serves the cause by suffering. The indignation of the country at the execution of Sydney and Russell alienated England finally and fatally from the House of Stuart. Lord Russell and his friend were canonized as the saints of the Revolution, but the harvest itself was gathered by statesmen of more common clay, yet perhaps better fitted for the working business of life.

Lord Russell's trial was attended with every feature which could concentrate the nation's attention upon it. The Duke of York was the actual and scarcely concealed prosecutor. Lady Rachel appeared in court as her husband's secretary. It is idle to say that he was unjustly convicted. He was privy to a scheme for armed resistance to the Government, and a Government which was afraid to punish him ought to have abdicated. Charles Stuart had been brought back by the deliberate will of the people. As long as he was on the throne he was entitled to defend both himself and his authority. Lord Russell was not, like Hampden, resisting an unconstitutional breach of the law. He was taking precautions against a danger which he anticipated, but which had not yet arisen. A government may be hateful, and we may admire the courage which takes arms against it; but the Government, while it exists, is not to be blamed for protecting itself with those weapons which the law places in his hands.

He died beautifully. Every effort was made to save him. His father pleaded his own exertions in bringing about the Restoration. But the Duke of York was inexorable, and Lord Russell was executed. The Earl was consoled after the Revolution with a dukedom. His mother, Lady Anne, did not live to recover from the shock of her son's death. In the midst of her wretchedness she found accidentally in a room in Woburn a pamphlet with an account of the Overbury murder. For the first time she learnt the dreadful story. She was found senseless, with her hand upon the open page, and never rallied from the blow.

Lady Rachel lived far into the following century, and was a venerable old lady before she rejoined her husband. Once at least while alive Lady Rachel visited Cheneys Chapel. Her foot had stood on the same stones where mine were standing; her eyes had rested on the same sculptured figures.

' I have accomplished it,' she wrote, ' and am none the worse for having satisfied my longing mind, and that is a little ease—such degree of it as I must look for. I had some business there, for that to me precious and delicious friend desired I would make a little monument for us, and I had never seen the place. I had set a day to see it with him not three months before he was carried thither, but was prevented by the boy's illness.'

' She would make a little monument.' And out of that modest hope of hers has grown the monstrous outrage upon taste and simplicity, which we may piously hope was neither designed nor approved by the admirable Lady Rachel.

Lord Russell had pressed his devotion to the cause of liberty beyond the law; another Russell had been accused of treason to the sacred traditions of the family. Edward, the youngest brother of the fourth Earl Francis, who lies with the rest at Cheneys, had a son, who was one of the few Russells that were famous in arms—the admiral who won the battle of La Hogue, saved England from invasion, and was rewarded with the Earldom of Orford. Admiral Russell, like Marlborough, notwithstanding his brilliant services, was beyond doubt in correspondence with the Court of St Germains, and equally beyond doubt held out hopes to the banished King that he might desert William and carry the fleet along with him. The real history of these mysterious transactions is unknown, and, perhaps, never will be known. William was personally unpopular. His manner was ungracious. He was guilty of the unpardonable sin of being a foreigner, which Englishmen could never forgive. A restoration like that of Charles II seemed at one time, at least, one of the chances which were on the cards—and cautious politicians may not have felt they were committing any serious violation of trust in learning directly from James the securities for rational liberty which he was ready to concede. The negotiation ended, however, in nothing—and it is equally likely that it was intended to end in nothing. James's own opinion was that ' Admiral Russell did but delude the King with the Prince of Orange's permission.' It is needless to speculate on the motives of conduct, which, if we knew them, we should be unable to enter into. To the student who looks back over the past, the element of uncertainty is eliminated. When the future, which to the living man is contingent and dim, obscuring his very duties to him, has become a realized fact, no effort of imagination will enable the subsequent inquirer to place himself in a position where the fact was but floating possibility. The services both of Churchill and Russell might be held great enough to save them from the censure of critics, who, in their arm-chairs at a distance of two centuries, moralize on the meannesses of great men.

The Admiral, at any rate, is not among his kindred in the Cheneys vault. He was buried at his own home, and his peerage and his lineage are extinct.

The Dukedom has made no difference in the attitude of the Bedford family. A more Olympian dignity has surrounded the chiefs of the house, but they have continued, without exception, staunch friends of liberty; advocates of the things called Reform and Progress, which have taken the place of the old Protestant cause; and the younger sons have fought gallantly like their forefathers in the front ranks of the battle. We may let the dukes glide by wearing the honours which democracy allows to stand, because they are gradually ceasing to have any particular meaning. We pass on to the last Russell for whom the vault at Cheneys has unlocked ' its marble jaws;' the old statesman who filled so large a place for half a century in English public life, whose whole existence from the time when he passed out of childhood was spent in sharp political conflict, under the eyes of the keenest party criticisms, and who carried his reputation off the stage at last, unspotted by a single act which his biographers were called on to palliate.

To the Tories, in the days of the Reform Bill, Lord John Russell was the tribune of an approaching violent revolution. To the Radicals he was the Moses who was leading the English nation into the promised land. The alarm and the hope were alike imaginary. The wave has gone by, the crown and peerage and church and primogeniture stand where they were, and the promised land, alas! is a land not running with corn and wine, but running only with rivers of gold, at which those who drink are not refreshed. To the enthusiasts of Progress the Reform

Bill of 1832 was to be a fountain of life, in which society was to renew its youth like the eagle. Highborn ignorance was to disappear from the great places of the nation; we were to be ruled only by Nature's aristocracy of genius and virtue; the inequalities of fortune were to be readjusted by a truer scale; and merit, and merit only, was to be the road to employment and distinction. We need not quarrel with a well-meant measure because foolish hopes were built upon it. But experienced men say that no one useful thing has been done by the Reformed Parliament which the old Parliament would have refused to do; and for the rest, it begins to be suspected that the reform of which we have heard so much is not the substitution of a wise and just government for a government which was not wise and just, but the abolishment of government altogether, and the leaving each individual man to follow what he calls his interest—a process under which the English people are becoming a congregation of contending atoms, scrambling every one of them to snatch a larger portion of good things than its fellow.

It is idle to quarrel with the inevitable. Each generation has its work to do. Old England could continue no longer; and the problem for the statesmen of the first half of this century was to make the process of transformation a quiet and not a violent one. The business of Lord John Russell was to save us from a second edition of the French Revolution; and if he thought that something higher or better would come of it than we have seen, or are likely to see, it is well that men are able to indulge in such pleasant illusions to make the road the lighter for them. The storms of his early life had long passed away before the end came. He remained the leader of the Liberal party in the House of Commons during the many years in which the administration was in the Liberal hands; and he played his part with a prudence and good sense, of which we have been more conscious, perhaps, since the late absence of these qualities. Lord John Russell (or Earl Russell as he became) never played with his country's interests for the advantage of his party. Calumny never whispered a suspicion either of his honour or his patriotism, and Tory and Radical alike followed him when he retired with affectionate respect. In Cheneys church there is no monument of him. His statue will stand appropriately in the lobby of the House, where he fought and won his many battles. It may be said of him, as was said of Peel, that we did not realize his worth till he was taken from us. In spite of progress, we have not produced another man who can make us forget his loss.

Here, too, beneath the stones, lies another pair, of whom the world spoke much, and knew but little—Earl Russell's young son, who died prematurely before his father, and that son's still younger wife. Lord Amberley also was a genuine Russell, full of talent, following truth and right wherever they seemed to lead him; and had life been allowed him he too would have left his mark on his generation. He was carried away, it was said, into extreme opinions. It is no unpardonable crime. His father, too, in his young days, had admired Napoleon and the French Revolution; had admired many things of which in age he formed a juster estimate. We do not augur well of the two-year-old colt whose paces are as sedate as those of an established roadster, who never rears when he is mounted, or flings out his heels in the overflow of heart and spirit. Our age has travelled fast and far in new ways, tossing off traditions old as the world as if they were no better than worn-out rags; and the ardent and hopeful Amberley galloped far in front in pursuit of what he called Liberty, not knowing that it was a false phantom which he was following; not freedom at all—but anarchy. The wise world held up its hands in horror; as if any man was ever good for anything whose enthusiasm in his youth has not outrun his understanding. Amberley, too, would have learnt his lesson had time been granted him. He would have learnt it in the best of schools— by his own experience. Happy those who have died young if they have left a name as little spotted as his with grosser faults and follies.

She, too, his companion, went along with him in his philosophy of progress, each most extravagant opinion tempting her to play with it. True and simple in herself, she had been bred in disdain of unreality. Transparent as air, pure as the fountain which bubbles up from below a glacier, she was encouraged by her very innocence in speculations against which a nature more earthly would have been on its guard. She so hated insincerity that in mere wantonness she trampled on affectation and conventiality, and she would take up and advocate theories which, if put in practice, would make society impossible, while she seemed to me as little touched by them herself as the seagull's wings are wetted when it plunges into the waves.

The singular ways of the two Amberleys were the world's wonder for a season or two. The world might as well have let them alone. The actual arrangements of things are so far from excellent that young ardent minds become Radical by instinct when they first become acquainted with the world as it actually is. Radicalism is tamed into reasonable limits when it has battered itself for a few years against the stubborn bars of fact, and the conversion is the easier when the Radical is the heir of an earldom. The Amberleys, who went farther than Lord Russell had ever done in the pursuit of imaginary Utopias, might have recoiled farther when they learnt that they were hunting after a dream. Peace be with them. They may dream on now, where the world's idle tattle can touch them no more.

The ghostly pageant of the Russells has vanished. The silent hours of the summer morning are past, and the sounds outside tell that the hamlet is awake and at its work. The quiet matron must resume the charge of the church keys, that intruders may not stray into the sanctuary unpermitted. In Catholic countries the church doors stand open; the peasant pauses on his way to the fields for a moment of meditation or a few words of prayer. The kneeling figures, on a week-day morning, are more impressive than Sunday rituals or preacher's homily. It was so once here in Cheneys, in the time of the poor priest whose figure is still on the wall. Was the Reformation, too, the chase of a phantom? The freedom of the church at all events is no longer permitted here in Protestant England. I, too, must go upon my way back to the village inn, where—for such things have to be remembered—breakfast and a young companion are waiting for me. It is worth while to spend a day at Cheneys, if only for the breakfast—breakfast on fresh pink trout from the Ches, fresh eggs, fresh yellow butter, cream undefiled by chalk, and home-made bread untouched with alum. The Russells have been the apostles of progress, but there is no progress in their own dominion. The ducal warranty is on everything which is consumed here.

The sun was shining an hour ago. It is now raining; it rained all yesterday; the clouds are coming up from the south and the wind is soft as oil. The day is still before us, and it is a day made for trout fishing. The chapel is not the only attraction at Cheneys. No river in England holds finer trout, nor trout more willing to be caught. Why fish will rise in one stream and not in another is a problem which we must wait to understand, as Bret Harte says, in 'another and a better world.' The Ches at any rate is one of the favoured waters. Great, too, is the Duke of Bedford—great in the millions he has spent on his tenants' cottages—great in the remission of his rents in the years when the seasons are unpropitious— great in the administration of his enormous property; but greater than all in the management of his fishing, for if he gives you leave to fish there, you have the stream for the day to yourself. You are in no danger of seeing your favourite pool already flogged by another sportsman, or of finding rows of figures before you fringing the river bank, waving their long wands in the air, each followed by his boy with basket and generally useless landing net. 'Competition' and 'the greatest happiness of the greatest number' are not heard of in this antique domain. A day's fishing at Cheneys means a day by the best water in England in the fisherman's paradise of solitude.

Such a day's privilege had been extended to me if I cared to avail myself of it, when I was coming down to see the chapel, and though my sporting days were over, and gun and rod had long lain undisturbed in their boxes, yet neither the art of fly-fishing, nor the enjoyment of it when once acquired and tasted, will leave us except with life. The hand does not forget its cunning, and opportunity begets the inclination to use it. I had brought my fishing case along with me. Shall I stay at the inn over the day and try what can be done? The rain and the prospect of another such breakfast decide it between them. The water-keeper is at the window— best of keepers—for he will accept a sandwich perhaps for luncheon, a pull from your flask, and a cigar out of your case, but other fee on no condition. The rain, he tells me, has raised the water, and the large fish are on the move, the May-fly has been down for two days. They were feeding on it last evening. If the sky clears they will take well in the afternoon; but the fly will not show till the rain stops.

The Cheneys fishing is divided in the middle by a mill. Below the mill the trout are in greatest numbers, but com- paratively small; above it is a long still deep pool where the huge monsters lie, and in common weather never stir till twilight. The keeper and I remember a summer evening some years ago, when at nightfall, after a burning day, the glittering surface of the water was dimpled with rings, and a fly thrown into the middle of these circles was answered more than once by a rush and scream of the reel; and a struggle which the darkness made more exciting. You may as well fish on the high road as in the mill-pool when the sun is above the horizon, and even at night you will rarely succeed there; but at the beginning of the May-fly season these large fish sometimes run up to the rapid stream at the pool head to feed. This the keeper decides shall be tried if the fly comes down. For the morning he will leave me to myself.

Does the reader care to hear of a day's fishing in a chalk stream fifteen miles from London? As music to the deaf, as poetry to the political economist, as a mountain landscape to the London cockney, so is a chalk stream trout-fishing to those who never felt their fingers tingle as the line whistles through the rings. For them I write no further; let them leave the page uncut and turn on to the next article.

Breakfast over, I start for the lower water. I have my boy with me home for the holidays. He carries the landing net, and we splash through the rain to the mill. The river runs for a quarter of a mile down under hanging bushes. As with other accomplishments when once learnt, eye and hand do the work in fly-fishing without reference to the mind for orders. The eye tells the hand how distant the bushes are, how near the casting line approaches them. If a gust of wind twists it into a heap, or sweeps it towards a dangerous bough, the wrist does something on the instant which sends the fly straight and unharmed into the water. Practice gives our different organs functions like the instinct of animals, who do what their habits require, yet know not what they do.

The small fish take freely—some go back into the water, the few in good condition into the basket, which, after a field or two, becomes perceptibly heavier. The governor, a small humble bee, used to be a good fly at Cheneys, and so did the black alder. Neither of them is of any use to-day. The season has been cold and late. The March brown answers best, with the never-failing red spinner. After running rapidly through two or three meadows, the river opens into a broad smooth shallow, where the trout are larger, and the water being extremely clear, are specially difficult to catch. In such a place as this, it is useless to throw your fly at random upon the stream. You must watch for a fish which is rising, and you must fish for him till you either catch him or disturb him. It is not enough to go below him and throw upwards, for though he lies with his head up-stream, his projecting eye looks back over his shoulders. You must hide behind a bunch of rushes. You must crawl along the grass with one arm only raised. If the sun is shining and the shadow of your rod glances over the gravel, you may get up and walk away. No fish within sight will stir then to the daintiest cast.

I see a fish close to the bank on the opposite side, lazily lifting his head as a fly floats past him. It is a long throw, but the wind is fair and he is worth an effort—once, twice, three times I fail to reach him. The fourth I land the fly on the far bank, and draw it gently off upon his very nose. He swirls in the water like a salmon as he sweeps round to seize it. There is a splash—a sharp jerk, telling unmistakably that something has given way. A large fish may break you honestly in weeds or round a rock or stump, and only fate is to blame, but to let yourself be broken on the first strike is unpardonable. What can have happened? Alas, the red-spinner has snapped in two at the turn—a new fly bought last week at ——'s, whose boast it has been that no fly of his was ever known to break or bend.

One grumbles on these occasions, for it is always the best fish which one loses; and as imagination is free, one may call him what weight one pleases. The damage is soon repaired. The basket fills fast as trout follows trout. It still rains, and I begin to think that I have had enough of it. I have promised to be at the mill at midday, and then we shall see.

Evidently the sky means mischief. Black thunderclouds pile up to windward, and heavy drops continue falling. But there is a break in the south as I walk back by the bank—a gleam of sunshine spans the valley with a rainbow, and an actual May-fly or two sails by which I see greedily swallowed. The keeper is waiting; he looks scornfully into my basket. Fish—did I call these herrings fish? I must try the upper water at all events. The large trout were feeding, but the fly was not yet properly on—we can have our luncheon first.

How pleasant is luncheon on mountain-side or river's bank, when you fling yourself down on fern or heather after your morning's work, and no daintiest *entrée* had ever such flavour as your sandwiches, and no champagne was ever so exquisite as the fresh stream water just tempered from your whisky flask. Then follows the smoke, when the keeper fills his pipe at your bag, and old adventures are talked over, and the conversation wanders on through anecdotes and experiences, till, as you listen to the shrewd sense and kindly feeling of your companion, you become aware that the steep difference which you had imagined to be created by education and habits of life had no existence save in your own conceit. Fortune is less unjust than she seems, and true hearts and clear-judging healthy minds are bred as easily in the cottage as the palace.

But time runs on, and I must hasten to the end of my story. The short respite from the wet is over. Down falls the rain again; rain not to be measured by inches, but by feet; rain such as has rarely been seen in England before this ' æstas mirabilis ' of 1879. It looks hopeless, but the distance by the road to the top of the water is not great. We complain if we are caught in a shower; we splash along in a deluge, in boots and waterproof, as composedly as if we were seals or otters. The river is rising and, as seldom happens with a chalk stream, it is growing discoloured. Every lane is running with a brown stream, which finds its way at last into the main channel. The highest point is soon reached. The first hundred yards are shallow, and to keep the cattle from straying a high iron railing runs along the bank. Well I knew that iron railing. You must stand on the lower bar to fish over it. If you hook a trout, you must play him from that uneasy perch in a rapid current among weeds and stones, and your attendant must use his landing net through the bars. Generally it is the liveliest spot in the river, but nothing can be done there to-day. There is a ford immediately above, into which the thick road-water is pouring, and the fish cannot see the fly. Shall we give it up? Not yet. Further down the mud settles a little, and by this time even the road has been washed clean, and less dirt comes off it. The flood stirs the trout into life and hunger, and their eyes, accustomed to the transparency of the chalk water, do not see you so quickly.

Below the shallow there is a pool made by a small weir, over which the flood is now rushing—on one side there is an open hatchway, with the stream pouring through. The banks are bushy, and over the deepest part of the pool the stem of a large ash projects into the river. Yesterday, when the water was lower, the keeper saw a four-pounder lying under that stem. Between the weir and the trees is an awkward spot, but difficulty is the charm of fly-fishing. The dangerous drop fly must be taken off; a drop fly is only fit for open water, where there is neither weed nor stump. The March brown is sent skimming at the tail of the casting line, to be dropped, if possible, just above the ash, and to be carried under it by the stream. It has been caught in a root, so it seems; or it is foul somewhere. Surely no fish ever gave so dead a pull. No; it is no root. The line shoots under the bank. There is a broad flash of white just below the surface, a moment's struggle, the rod springs straight, and the line comes back unbroken. The March brown is still floating at the end of it. It was a big fish, perhaps the keeper's very big one; he must have been lightly hooked, and have rubbed the fly out of his mouth.

But let us look closer. The red spinner had played false in the morning; may not something like it have befallen the March brown? Something like it, indeed. The hook has straightened out as if, instead of steel, it had been made of copper. A pretty business! I try another, and another, with the same result. The heavy trout take them, and one bends and the next breaks. Oh! ————! Well for Charles Kingsley that he had gone before he heard of a treason which would have broken his trust in man. You, in whose praise I have heard him so often eloquent! You who never dealt in shoddy goods. You who were faithful if all else were faithless, and redeemed the credit of English tradesmen! You had not then been in the school of progress and learnt that it was the buyer's business to distinguish good from bad. You never furnished your customers with cheap and nasty wares, fair-looking to the eye and worthless to the touch and trial. In those days you dealt with gentlemen, and you felt and traded like a gentleman yourself. And now you too, have gone the way of your fellows. You are making a fortune, as you call it, out of the reputation which you won honourably in better days. You have given yourself over to competition and semblance. You have entered for the race among the sharpers and will win by knavery and tricks like the rest. I will not name you for the sake of the old times, when C. K. and I could send you a description of a fly from the farthest corner of Ireland, and by return of post would come a packet tied on hooks which Kendal and Limerick might equal, but could not excel. You may live on undenounced for me; but read C. K.'s books over again; repent of your sins, go back to honest ways, and renounce the new gospel in which whosoever believes shall not be saved.

But what is to be done? Spite of the rain the river is now covered with drowned May-flies, and the trout are taking them all round. I have new May-flies from the same quarter in my book, but it will be mere vexation to try them. Luckily for me there are a few old ones surviving from other days. The gut is brown with age—but I must venture it. If this breaks I will go home, lock away my rod, and write an essay on the effects of the substitution of Political Economy for the Christian faith.

On, then, goes one of these old flies. It looks well. It bears a mild strain, and, like Don Quixote with his helmet, I will not put it to a severe trial. Out it shoots over the pool, so natural looking that I cannot distinguish it from a real fly which floats at its side. I cannot, nor can that large trout in the smooth water above the fall. He takes it, springs into the air, and then darts at the weir to throw himself over. If he goes down he is lost. Hold on. He has the stream to help him, and not an inch of line can be spared. The rod bends double, but the old gut is true. Down the fall he is not to go. He turns up the pool, he makes a dart for the hatchway,—but if you can stand a trout's first rush you need not fear him in fair water afterwards. A few more efforts and he is in the net and on the bank, not the keeper's four-pounder, but a handsome fish, which I know that he will approve.

He had walked down the bank pensively while I was in the difficulty with my flies, meditating, perhaps, on idle gentlemen, and reflecting that if the tradesmen were knaves the gentlemen were correspondingly fools. He called to me to come to him just as I had landed my trout. He was standing by the side of the rapid stream at the head of the mill pool. It was as he had foretold; the great fish had come up, and were rolling like salmon on the top of the water gulping down the May-flies. Even when they are thus carelessly ravenous, the clearness of the river creates a certain difficulty in catching them in ordinary times, but to-day the flood made caution superfluous. They were splashing on the surface close to our feet, rolling about in a negligent gluttony which seemed to take from them every thought of danger, for a distance of at least three hundred yards.

There was no longer any alarm for the tackle, and it was but to throw the fly upon the river, near or far, for a trout instantly to seize it. There was no shy rising where suspicion balks the appetite. The fish were swallowing with a deliberate seriousness every fly which drifted in their reach, snapping their jaws upon it with a gulp of satisfaction. The only difficulty was in playing them when hooked with a delicate chalk-stream casting-line. For an hour and a half it lasted, such an hour and a half of trout fishing as I had never seen and shall never see again. The ease of success at last became wearisome. Two large baskets were filled to the brim. Accident had thrown in my way a singular opportunity which it would have been wrong to abuse, so I decided to stop. We emptied out our spoils upon the grass, and the old keeper said that long as he had known the river he had never but once seen so many fish of so large size taken in the Ches in a single day by a single rod.

How can a reasonable creature find pleasure in having performed such an exploit? If trout were wanted for human food, a net would have answered the purpose with less trouble to the man and less annoyance to the fish. Throughout creation man is the only animal—man, and the dogs and cats which have learnt from him—who kills, for the sake of killing, what he does not want, and calls it sport. All other animals seize their prey only when hungry, and are satisfied when their hunger is appeased.

Such, it can only be answered, is man's disposition. He is a curiously formed creature, and the appetite for sport does not seem to disappear with civilization. The savage in his natural state hunts, as the animals hunt, to support his life; the sense of sport is strongest in the elaborately educated and civilized. It may be that the taste will die out before 'Progress.' Our descendants perhaps, a few generations hence, may look back upon a pheasant battue as we look back on bear-baiting and bull-fighting, and our mild offspring, instructed in the theory of development, may see a proof in their fathers' habits that they come of a race who were once crueller than tigers, and will congratulate themselves on the change. So they will think, if they judge us as we judge our forefathers of the days of the Plantagenets and Tudors, and both we and they may be perhaps mistaken. Half the lives of men in mediæval Europe was spent in fighting. Yet from mediæval Europe came the knightly graces of courtesy and chivalry. The modern soldier, whose trade is war, yet hates and dreads war more than civilians dread it. The sportsman's knowledge of the habits of animals gives him a kindly feeling towards them notwithstanding, and sporting tends rather to their preservation than their destruction. The human race may become at last vegetarians and water-drinkers. Astræa may come back, and man may cease to take the life of bird, or beast, or fish. But the lion will not lie down with the lamb, for lambs and lions will no longer be; the eagle will not feed beside the dove, for doves will not be allowed to consume grain which might have served as human food, and will be extinct as the dodo. It may be all right and fit and proper : a world of harmless vegetarians may be the appropriate outcome of the development of humanity. But we who have been born in a ruder age do not aspire to rise beyond the level of our own times. We have toiled, we have suffered, we have enjoyed, as the nature which we have received has prompted us. We blame our fathers' habits; our children may blame ours in turn; yet we may be sitting in judgment, both of us, on matters of which we know nothing.

The storm has passed away, the dripping trees are sparkling in the warm and watery sunset. Back, then, to our inn, where dinner waits for us, the choicest of our own trout, pink as salmon, with the milky curd on them, and no sauce to spoil the delicacy of their flavour. Then bed, with its lavender-scented sheets and white curtains, and sleep, sound sweet sleep, that loves the country village and comes not near a London bedroom. In the morning, adieu to Cheneys, with its red gable-ends and chimneys, its venerable trees, its old-world manners, and the solemn memories of its mausoleum. Adieu, too, to the river, which, ' though men may come and men may go,' has flowed and will flow on for ever, winding among its reed beds, murmuring over its gravelly fords, heedless of royal dynasties, uncaring whether Cheney or Russell calls himself lord of its waters, graciously turning the pleasant corn mills in its course, unpolluted by the fetid refuse of manufactures, and travelling on to the ocean bright and pure and uncharged with poison, as in the old times when the priest sung mass in the church upon the hill and the sweet soft matins bell woke the hamlet to its morning prayers.

JOHN HENRY NEWMAN[1]

My dear ———. My present letter will be given to a single figure. When I entered at Oxford, John Henry Newman was beginning to be famous. The responsible authorities were watching him with anxiety; clever men were looking with interest and curiosity on the apparition among them of one of those persons of indisputable genius who was likely to make a mark upon his time. His appearance was striking. He was above the middle height, slight and spare. His head was large, his face remarkably like that of Julius Cæsar. The forehead, the shape of the ears and nose, were almost the same. The lines of the mouth were very peculiar, and I should say exactly the same. I have often thought of the resemblance, and believed that it extended to the temperament. In both there was an original force of character which refused to be moulded by circumstances, which was to make its own way, and become a power in the world; a clearness of intellectual perception, a disdain for conventionalities, a temper imperious and wilful, but along with it a most attaching gentleness, sweetness, singleness of heart and purpose. Both were formed by nature to command others, both had the faculty of attracting to themselves the passionate devotion of their friends and followers, and in both cases, too, perhaps the devotion was rather due to the personal ascendency of the leader than to the cause which he represented. It was Cæsar, not the principle of the empire, which overthrew Pompey and the constitution. *Credo in Newmannum* was a common phrase at Oxford, and is still unconsciously the faith of nine-tenths of the English converts to Rome.

When I first saw him he had written his book upon the Arians. An accidental application had set him upon it, at a time, I believe, when he had half resolved to give himself to

[1] [Letter III.]

230

science and mathematics, and had so determined him into a theological career. He had published a volume or two of parochial sermons. A few short poems of his had also appeared in the ' British Magazine ' under the signature of ' Delta,' which were reprinted in the ' Lyra Apostolica.' They were unlike any other religious poetry which was then extant. It was hard to say why they were so fascinating. They had none of the musical grace of the ' Christian Year.' They were not harmonious; the metre halted, the rhymes were irregular, yet there was something in them which seized the attention, and would not let it go. Keble's verses flowed in soft cadence over the mind, delightful, as sweet sounds are delightful, but are forgotten as the vibrations die away. Newman's had pierced into the heart and mind, and there remained. The literary critics of the day were puzzled. They saw that he was not an ordinary man; what sort of an extraordinary man he was they could not tell. ' The eye of Melpomene has been cast upon him,' said the omniscient (I think) ' Athenæum ';[2] ' but the glance was not fixed or steady.' The eye of Melpomene had extremely little to do in the matter. Here were thoughts like no other man's thoughts, and emotions like no other man's emotions. Here was a man who really believed his creed, and let it follow him into all his observations upon outward things. He had been travelling in Greece; he had carried with him his recollections of Thucydides, and while his companions were sketching olive gardens and old castles and picturesque harbours at Corfu, Newman was recalling the scenes which those harbours had witnessed thousands of years ago in the civil wars which the Greek historian had made immortal. There was nothing in this that was unusual. Any one with a well-stored memory is affected by historical scenery. But Newman was oppressed with the sense that the men who had fallen in that desperate strife were still alive, as much as he and his friends were alive.

[2] Perhaps it was not the *Athenæum*. I quote from memory. I remember the passage from the amusement which it gave me; but it was between forty and fifty years ago, and I have never seen it since.

> *Their spirits live in awful singleness,*

he says,

> *Each in its self-formed sphere of light or gloom.*

We should all, perhaps, have acknowledged this in words.
It is happy for us that we do not all realize what the words
mean. The minds of most of us would break under the strain.

Other conventional beliefs, too, were quickened into startling realities. We had been hearing much in those days about
the benevolence of the Supreme Being, and our corresponding
obligation to charity and philanthropy. If the received creed
was true, benevolence was by no means the only characteristic
of that Being. What God loved we might love; but there
were things which God did not love; accordingly we found
Newman saying to us—

> *Christian, would'st thou learn to love?*
> *First learn thee how to hate.*
>
> .   .   .   .   .
>
> *Hatred of sin and zeal and fear*
> *Lead up the Holy Hill;*
> *Track them, till charity appear*
> *A self-denial still.*

It was not austerity that made him speak so. No one was
more essentially tender-hearted. But he took the usually
accepted Christian account of man and his destiny to be
literally true, and the terrible character of it weighed upon
him.

> *Sunt lacrymæ rerum et mentem mortalia tangunt.*

He could be gentle enough in other moods. 'Lead, kindly
Light,' is the most popular hymn in the language. All of us,
Catholic, Protestant, or such as can see their way to no positive
creed at all, can here meet on common ground and join in a
common prayer. Familiar as the lines are, they may here be
written down once more :—

> *Lead, kindly Light, amid the encircling gloom*
> *Lead Thou me on.*
> *The night is dark, and I am far from home,*
> *Lead thou me on.*
> *Keep Thou my feet; I do not ask to see*
> *Far distant scenes—one step enough for me.*

I was not ever thus, nor prayed that Thou
   Should'st lead me on.
I loved to choose and see my path; but now
   Lead Thou me on.
I loved the garish day, and, spite of fears,
Pride ruled my will. Remember not past years.

So long Thy power hath blest us, sure it will
   Still lead us on,
O'er moor and fen, o'er crag and torrent, till
   The night is gone,
And with the morn those angel faces smile
Which I have loved long since, and lost awhile.

It has been said that men of letters are either much less
or much greater than their writings. Cleverness and the
skilful use of other people's thoughts produce works which
take us in till we see the authors, and then we are disenchanted.
A man of genius, on the other hand, is a spring in which
there is always more behind than flows from it. The painting
or the poem is but a part of him inadequately realized, and his
nature expresses itself, with equal or fuller completeness, in
his life, his conversation, and personal presence. This was
eminently true of Newman. Greatly as his poetry had struck
me, he was himself all that the poetry was, and something
far beyond. I had then never seen so impressive a person.
I met him now and then in private; I attended his church and
heard him preach Sunday after Sunday; he is supposed to have
been insidious, to have led his disciples on to conclusions to
which he designed to bring them, while his purpose was
carefully veiled. He was, on the contrary, the most transparent
of men. He told us what he believed to be true. No one who
has ever risen to any great height in this world refuses to move
till he knows where he is going. He is impelled in each step
which he takes by a force within himself. He satisfies himself
only that the step is a right one, and he leaves the rest
to Providence. Newman's mind was world-wide. He was
interested in everything which was going on in science, in
politics, in literature. Nothing was too large for him, nothing
too trivial, if it threw light upon the central question, what

man really was, and what was his destiny. He was careless about his personal prospects. He had no ambition to make a career, or to rise to rank and power. Still less had pleasure any seductions for him. His natural temperament was bright and light; his senses, even the commonest, were exceptionally delicate. I was told that, though he rarely drank wine, he was trusted to choose the vintages for the college cellar. He could admire enthusiastically any greatness of action and character, however remote the sphere of it from his own. Gurwood's ' Dispatches of the Duke of Wellington ' came out just then. Newman had been reading the book, and a friend asked him what he thought of it. ' Think?' he said, ' it makes one burn to have been a soldier.' But his own subject was the absorbing interest with him. Where Christianity is a real belief, where there are distinct convictions that a man's own self and the millions of human beings who are playing on the earth's surface are the objects of a supernatural dispensation, and are on the road to heaven or hell, the most powerful mind may well be startled at the aspect of things. If Christianity was true, since Christianity was true (for Newman at no time doubted the reality of the revelation), then modern England, modern Europe, with its march of intellect and its useful knowledge and its material progress, was advancing with a light heart into ominous conditions. Keble had looked into no lines of thought but his own. Newman had read omnivorously; he had studied modern thought and modern life in all its forms, and with its many-coloured passions. He knew, of course, that many men of learning and ability believed that Christianity was not a revelation at all, but had been thrown out, like other creeds, in the growth of the human mind. He knew that doubts of this kind were inevitable results of free discussion and free toleration of differences of opinion; and he was too candid to attribute such doubts, as others did, to wickedness of heart. He could not, being what he was, acquiesce in the established religion as he would acquiesce in the law of the land, because it was there, and because the country had accepted it, and because good general reasons could be given for assuming it to be right. The soundest arguments, even the arguments of Bishop Butler himself, went no farther than to establish a probability. But religion with Newman was a personal thing between himself and his Maker, and it was not possible to feel love and devotion to a Being whose existence was merely probable. As Carlyle says of himself when in a similar condition, a religion which was not a certainty was a mockery and a horror; and unshaken and unshakable as his own convictions were, Newman evidently was early at a loss for the intellectual grounds on which the claims of Christianity to abstract belief could be based. The Protestant was satisfied with the Bible, the original text of which, and perhaps the English translation, he regarded as inspired. But the inspiration itself was an assumption, and had to be proved; and Newman, though he believed the inspiration, seems to have recognized earlier than most of his contemporaries that the Bible was not a single book, but a national literature, produced at intervals, during many hundred years, and under endless varieties of circumstances. Protestant and Catholic alike appealed to it, and they could not both be right. Yet if the differences between them were essential, there must be some authority capable of deciding between them. The Anglican Church had a special theology of its own, professing to be based on the Bible. Yet to suppose that each individual left to himself would gather out of the Bible, if able and conscientious, exactly these opinions and no others, was absurd and contrary to experience. There were the creeds; but on what authority did the creeds rest? On the four councils? or on other councils, and, if other, on which? Was it on the Church? and, if so, on what Church? The Church of the Fathers? or the Church still present and alive and speaking? If for living men, among whom new questions were perpetually rising, a Church which was also living could not be dispensed with, then what was that Church, and to what conclusions would such an admission lead us?

With us undergraduates Newman, of course, did not enter on such important questions, although they were in the air, and we talked about them among ourselves. He, when we met him, spoke to us about subjects of the day, of literature, of public persons and incidents, of everything which was generally interesting. He seemed always to be better informed on common topics of conversation than any one else who was present. He was never condescending with us, never didactic or authoritative; but what he said carried conviction along with it. When we were wrong he knew why we were wrong, and excused our mistakes to ourselves while he set us right. Perhaps his supreme merit as a talker was that he never tried to be witty or to say striking things. Ironical he could be, but not ill-natured. Not a malicious anecdote was ever heard from him. Prosy he could not be. He was lightness itself—the lightness of elastic strength—and he was interesting because he never talked for talking's sake, but because he had something real to say.

Thus it was that we, who had never seen such another man, and to whom he appeared, perhaps, at special advantage in contrast with the normal college don, came to regard Newman with the affection of pupils (though pupils, strictly speaking, he had none) for an idolized master. The simplest word which dropped from him was treasured as if it had been an intellectual diamond. For hundreds of young men *Credo in Newmannum* was the genuine symbol of faith.

Personal admiration, of course, inclined us to look to him as a guide in matters of religion. No one who heard his sermons in those days can ever forget them. They were seldom directly theological. We had theology enough and to spare from the select preachers before the university. Newman, taking some Scripture character for a text, spoke to us about ourselves, our temptations, our experiences. His illustrations were inexhaustible. He seemed to be addressing the most secret consciousness of each of us—as the eyes of a portrait appear to look at every person in a room. He never exaggerated; he was never unreal. A sermon from him was a poem, formed on a distinct idea, fascinating by its subtlety, welcome—how welcome!—from its sincerity, interesting from its originality, even to those who were careless of religion; and to others who wished to be religious, but had found religion dry and wearisome, it was like the springing of a fountain out of the rock.

The hearts of men vibrate in answer to one another like the strings of musical instruments. These sermons were, I suppose, the records of Newman's own mental experience. They appear to me to be the outcome of continued meditation upon his fellow-creatures and their position in this world; their awful responsibilities; the mystery of their nature, strangely mixed of good and evil, of strength and weakness. A tone, not of fear, but of infinite pity runs through them all, and along with it a resolution to look facts in the face; not to fly to evasive generalities about infinite mercy and benevolence, but to examine what revelation really has added to our knowledge, either of what we are or of what lies before us. We were met on all sides with difficulties; for experience did not confirm, it rather contradicted, what revelation appeared distinctly to assert. I recollect a sermon from him—I think in the year 1839—I have never read it since; I may not now remember the exact words, but the impression left is ineffaceable. It was on the trials of faith, of which he gave different illustrations. He supposed, first, two children to be educated together, of similar temperament and under similar conditions, one of whom was baptized and the other unbaptized. He represented them as growing up equally amiable, equally upright, equally reverent and God-fearing, with no outward evidence that one was in a different spiritual condition from the other; yet we were required to believe, not only that their condition was totally different, but that one was a child of God, and his companion was not.

Again, he drew a sketch of the average men and women who made up society, whom we ourselves encountered in daily life, or were connected with, or read about in newspapers. They were neither special saints nor special sinners. Religious men had faults, and often serious ones. Men careless of religion were often amiable in private life—good husbands, good fathers, steady friends, in public honourable, brave, and patriotic. Even in the worst and wickedest, in a witch of Endor, there was a human heart and human tenderness. None seemed good enough for heaven, none so bad to deserve to be consigned to the company of evil spirits, and to remain in pain and misery for ever. Yet all these people were, in fact, divided one from the other by an invisible line of separation. If they were to die on the spot as they actually were, some would be saved, the rest would be lost—the saved to have eternity of happiness, the lost to be with the devils of hell.

Again, I am not sure whether it was on the same occasion, but it was in following the same line of thought, Newman described closely some of the incidents of our Lord's passion; he then paused. For a few moments there was a breathless silence. Then, in a low, clear voice, of which the faintest vibration was audible in the farthest corner of St Mary's, he said, ' Now, I bid you recollect that He to whom these things were done was Almighty God.' It was as if an electric stroke had gone through the church, as if every person present understood for the first time the meaning of what he had all his life been saying. I suppose it was an epoch in the mental history of more than one of my Oxford contemporaries.

Another sermon left its mark upon me. It was upon evidence. I had supposed up to that time that the chief events related to the Gospels were as well authenticated as any other facts of history. I had read Paley and Grotius at school, and their arguments had been completely satisfactory to me. The Gospels had been written by apostles or companions of apostles. There was sufficient evidence, in Paley's words, ' that many professing to be original witnesses of the Christian miracles had passed their lives in labours, dangers, and sufferings in attestation of the accounts which they delivered.' St Paul was a further and independent authority. It was not conceivable that such men as St Paul and the other apostles evidently were should have conspired to impose a falsehood upon the world, and should have succeeded in doing it undetected in an age exceptionally cultivated and sceptical. Gibbon I had studied also, and had thought about the five causes by which he explained how Christianity came to be believed; but they had seemed to me totally inadequate. I was something more than surprised, therefore, when I heard Newman say that Hume's argument against the credibility of miracles was logically sound. The laws of nature, so far as could be observed, were uniform, and in any given instance it *was* more likely, as a mere matter of evidence, that men should deceive or be deceived, than that those laws should have been deviated from. Of course he did not leave the matter in this position. Hume goes on to say that he is speaking of evidence as addressed to the reason; the Christian religion addresses itself to faith, and the credibility of it is therefore unaffected by his objection. What Hume said in irony Newman accepted in earnest. Historically the proofs were insufficient, or sufficient only to create a sense of probability. Christianity was apprehended by a faculty essentially different. It was called faith. But what was faith, and on what did it rest? Was it as if mankind had been born with but four senses, by which to form their notions of things external to them, and that a fifth sense of sight was suddenly conferred on favoured individuals, which converted conjecture into certainty? I could not tell. For myself this way of putting the matter gave me no new sense at all, and only taught me to distrust my old ones.

I say at once that I think it was injudicious of Newman to throw out before us thus abruptly an opinion so extremely agitating. I explain it by supposing that here, as elsewhere, his sermons contained simply the workings of his own mind, and were a sort of public confession which he made as he went along. I suppose that something of this kind had been passing through him. He was in advance of his time. He had studied the early fathers; he had studied Church history, and the lives of the saints and martyrs. He knew that the hard and fast line which Protestants had drawn at which miracles had ceased was one which no historical canon could reasonably defend. Stories of the exercise of supernatural power ran steadily from the beginning to the latest period of the Church's existence; many of them were as well supported by evidence as the miracles of the New Testament; and if reason was to be the judge, no arbitrary separation of the age of the Apostles from the age of their successors was possible. Some of these stories might be inventions, or had no adequate authority for them; but for others there was authority of eye-witnesses; and if these were to be set aside by a peremptory act of will as unworthy of credit, the Gospel miracles themselves might fall before the same methods. The argument of Hume was already silently applied to the entire post-apostolic period. It had been checked by the traditionary reverence for the

Bible. But this was not reason; it was faith. Perhaps, too, he saw that the alternative did not lie as sharply as Paley supposed, between authentic fact and deliberate fraud. Legends might grow; they grew every day, about common things and persons, without intention to deceive. Imagination, emotion, affection, or, on the other side, fear and animosity, are busy with the histories of men who have played a remarkable part in the world. Great historic figures—a William Tell, for instance—have probably had no historical existence at all, and yet are fastened indelibly into national traditions. Such reflections as these would make it evident that if the Christian miracles were to be believed, not as possibly or probably true, but as indisputably true—true in such a sense that a man's life on earth, and his hope for the future, could be securely based upon them—the history must be guaranteed by authority different in kind from the mere testimony to be gathered out of books. I suppose every thinking person would now acknowledge this to be true. And we see, in fact, that Christians of various persuasions supplement the evidence in several ways. Some assume the verbal inspiration of the Bible; others are conscious of personal experiences which make doubt impossible. Others, again, appeal justly to the existence of Christianity as a fact, and to the power which it has exerted in elevating and humanizing mankind. Newman found what he wanted in the living authority of the Church, in the existence of an organized body which had been instituted by our Lord Himself, and was still actively present among us as a living witness of the truth. Thus the imperfection of the outward evidence was itself an argument for the Catholic theory. All religious people were agreed that the facts of the Gospel narrative really happened as they were said to have happened. Proof there must be somewhere to justify the conviction; and proof could only be found in the admission that the Church, the organized Church with its bishops and priests, was not a human institution, but was the living body through which the Founder of Christianity Himself was speaking to us.

Such, evidently, was one use to which Hume's objection could be applied, and to those who, like Newman, were provided with the antidote, there was no danger in admitting the force of it. Nor would the risk have been great with his hearers if they had been playing with the question as a dialectical exercise. But he had made them feel and think seriously about it by his own intense earnestness, and brought up as most of them had been to believe that Christianity had sufficient historical evidence for it, to be suddenly told that the famous argument against miracles was logically valid after all, was at least startling. The Church theory, as making good a testimony otherwise defective, was new to most of us, and not very readily taken in. To remove the foundation of a belief, and to substitute another, is like putting new foundations to a house—the house itself may easily be overthrown in the process. I have said before that in a healthy state of things religion is considered too sacred to be argued about. It is believed as a matter of duty, and the why or the wherefore are not so much as thought about. Revolutions are not far off when men begin to ask whence the sovereign derives his authority. Scepticism is not far off when they ask why they believe their creed. We had all been satisfied about the Gospel history; not a shadow of doubt had crossed the minds of one of us; and though we might not have been able to give a logical reason for our certitude, the certitude was in us, and might well have been let alone. I afterwards read Hume attentively, and though old associations prevented me from recognizing the full force of what he had to say, no doubt I was unconsciously affected by him. I remember insisting to a friend that the essential part of religion was morality. My friend replied that morality was only possible to persons who received power through faith to keep the commandments. But this did not satisfy me, for it seemed contrary to fact. There were persons of great excellence whose spiritual beliefs were utterly different. I could not bring myself to admit that the goodness, for instance, of a Unitarian was only apparent. After all is said, the visible conduct of men is the best test that we can have of their inward condition. If not the best, where are we to find a better?

TRACT XC AND ITS CONSEQUENCES[1]

My dear ——. After I had taken my degree, and before I re-entered upon residence as fellow, my confidence in my Oxford teachers underwent a further trial. I spent some months in Ireland in the family of an Evangelical clergyman.[2] I need not mention names which have no historical notability. My new friends were favourable specimens of a type which was then common in Ireland. The Church of England was becoming semi-Catholic. The Church of Ireland left Catholicism to those to whom it properly belonged. It represented the principles of the Reformation. It was a branch of what Mr Gladstone has called the Upas-tree of Protestant ascendency. Mr —— and the circle into which I was thrown were, to begin with, high-bred and cultivated gentlemen. They had seen the world. Some of them had been connected with the public movements of the time. O'Connell was then in his glory. I heard Irish affairs talked of by those who lived in the midst of them. A sharp line of division among the people distinguished the Protestants from the Catholics. The Protestants were industrious and thriving. Mendicancy, squalor, and misery went along with the flocks of the priest, whether as cause or effect of their belief, or in accidental connection with it, I could not tell. The country was outwardly quiet, but there were ominous undertones of disaffection. There were murders now and then in the mountains, and I was startled at the calmness with which they were spoken of. We were in the midst of the traditions of 1798. My friend's father had been attacked in his palace, and the folios in the library bore marks of having been used to barricade the windows. He himself spoke as if he was living on a volcano; but he was as unconcerned as a soldier at his post, and so far as outward affairs went he was as kind to Catholics as to Protestants. His outdoor servants were Catholics, and they seemed attached

[1] [Letter IV.]　　　　　　　　　[2] [In 1842.]

242

to him; but he knew that they belonged to secret societies, and that if they were ordered to kill him they would do it. The presence of exceptional danger elevates characters which it does not demoralize. There was a quiet good sense, an intellectual breadth of feeling in this household, which to me, who had been bred up to despise Evangelicals as unreal and affected, was a startling surprise. I had looked down on Dissenters especially, as being vulgar among their other enormities; here were persons whose creed differed little from that of the Calvinistic Methodists, yet they were easy, natural, and dignified. In Ireland they were part of a missionary garrison, and in their daily lives they carried the colours of their faith. In Oxford, reserve was considered a becoming feature in the religious character. The doctrines of Christianity were mysteries, and mysteries were not to be lightly spoken of. Christianity at —— was part of the atmosphere which we breathed; it was the great fact of our existence, to which everything else was subordinated. Mystery it might be, but not more of a mystery than our own bodily lives, and the system of which we were a part. The problem was to arrange all our thoughts and requirements in harmony with the Christian revelation, and to act it out consistently in all that we said and did. The family devotions were long, but there was no formalism, and everybody took a part in them. A chapter was read and talked over, and practical lessons were drawn out of it; otherwise there were no long faces or solemn affectations; the conversations were never foolish or trivial; serious subjects were lighted up as if by an ever-present spiritual sunshine.

Such was the new element into which I was introduced under the shadow of the Irish Upas-tree; the same uniform tone being visible in parents, in children, in the indoor servants, and in the surrounding society. And this was Protestantism. This was the fruit of the Reformation which we had been learning at Oxford to hate as rebellion and to despise as a system without foundation. The foundation of it was faith in the authority of Holy Scripture, which was supposed to be verbally inspired; and as a living witness, the presence of Christ in the heart. Here, too, the letter of the word was allowed to require a living authentication. The Anglo-Catholics at Oxford maintained that Christ was present in the Church; the Evangelicals said that he was present in the individual believing soul, and why might they not be right? So far as Scripture went they had promises to allege for themselves more definite than the Catholics. If the test was personal holiness, I for my own part had never yet fallen in with any human beings in whose actions and conversation the spirit of Christ was more visibly present.

My feelings of reverence for the Reformers revived. Fact itself was speaking for them. Beautiful pictures had been put before us of the mediæval Church which a sacrilegious hand had ruthlessly violated. Here on one side we saw the mediæval creed in full vitality with its fruits upon it which our senses could test; on the other, equally active, the fruits of the teaching of Luther and Calvin. I felt that I had been taken in, and I resented it. Modern history resumed its traditionary English aspect. I went again over the ground of the sixteenth century. Unless the intelligent part of Europe had combined to misrepresent the entire period, the corruption of Roman Catholicism had become intolerable. Put the matter as the Roman Catholics would, it was a fact impossible to deny, that they had alienated half Europe, that the Teutonic nations had risen against them in indignation, and had substituted for the Christianity of Rome the Christianity of the Bible. They had tried, and tried in vain, to extinguish the revolt in blood, and the national life of modern England had grown up out of their overthrow. With the Anglo-Catholics the phenomena were the same in a lighter form. The Anglo-Catholics, too, had persecuted as far as they dared; they, too, had been narrow, cruel, and exclusive. Peace and progress had only been made possible when their teeth were drawn and their nails pared, and they were tied fast under the control of Parliament. History, like present reality, was all in favour of the views of my Evangelical friends.

And if history was in their favour, so were analogy and general probability. Mediæval theology had been formed at a time when the relations of matter and spirit had been guessed at by imagination, rather than studied with care and observation. Mind it was now known could only act on matter through the body specially attached to it. Ideas reached the mind through the senses, but it was by method and sequence which, so far as experience went, was never departed from. The Middle Ages, on the other hand, believed in witchcraft and magic. Incantation could call up evil angels and control the elements. The Catholic theory of the Sacraments was the counterpart of enchantment. Outward mechanical acts which, except as symbols, had no meaning, were supposed to produce spiritual changes, and spoken words to produce, like spells, changes in material substance. The imposition of a bishop's hands conferred supernatural powers. An ordained priest altered the nature of the elements in the Eucharist by consecrating them. Water and a prescribed formula regenerated an infant in baptism. The whole Church, it was true, had held these opinions down to the sixteenth century. But so it had believed that medicine was only efficacious if it was blessed; so it had believed that saints' relics worked miracles. Larger knowledge had taught us that magic was an illusion, that spells and charms were fraud or folly. The Reformers in the same way had thrown off the notion that there was anything mysterious or supernatural in the clergy or the Sacraments. The clergy in their opinion were like other men, and were simply set apart for the office of teaching the truths of religion. The Sacraments were symbols, which affected the moral nature of those who could understand them, as words or pictures, or music, or anything else which had an intelligible spiritual meaning. They brought before the mind in a lively manner the facts and principles of Christianity. To regard them as more was superstition and materialism. Evangelicalism had been represented to me as weak and illiterate. I had found it in harmony with reason and experience, and recommended as it was by personal holiness in its professors, and general beauty of mind and character, I concluded that Protestantism had more to say for itself than my Oxford teachers had allowed.

For the first time, too, among these good people I was

introduced to Evangelical literature. Newton[3] and Faber[4] had given me good reasons when I was a boy for believing the Pope to be the man of sin : but I had read nothing of Evangelical positive theology, and books like the ' Pilgrim's Progress ' were nothing less than a revelation to me. I do not mean that I could adopt the doctrine in the precise shape in which it was represented to me, that I was *converted,* or anything of that kind; but I perceived that persons who rejected altogether the theory of Christianity which I had been taught to regard as the only tenable one, were as full of the spirit of Christ, and had gone through as many, as various, and as subtle Christian experiences as the most developed saint in the Catholic calendar. I saw it in their sermons, in their hymns, in their conversation. A clergyman, who was afterwards a bishop in the Irish Church, declared in my hearing that the theory of a Christian priesthood was a fiction; that the notion of the Sacraments as having a mechanical efficacy irrespective of their conscious effect upon the mind of the receiver was an idolatrous superstition; that the Church was a human institution, which had varied in form in different ages, and might vary again; that it was always fallible; that it might have bishops in England, and dispense with bishops in Scotland and Germany; that a bishop was merely an officer; that the apostolical succession was probably false as a fact—and, if a fact, implied nothing but historical continuity. Yet the man who said these things had devoted his whole life to his Master's service—thought of nothing else, and cared for nothing else.

The opinions were of no importance in themselves; I was, of course, aware that many people held them; but I realized now for the first time that clergymen of weight and learning in the Church of England, ordained and included in its formularies, could think in this way and openly say so, and

[3] [The reference may be either to the Rev. John Newton (1725-1807), the friend of Cowper, or to Dr. Robert Newton (1780-1854), a Wesleyan minister, whose sermons and pamphlets had a great contemporary influence.]

[4] [George Stanley Faber (1773-1854), an evangelical clergyman whose numerous writings include criticism of Romanism.]

that the Church to which Newman and Keble had taught us to look as our guide did not condemn them. Clearly, therefore, if the Church equally admitted persons who held the sacramental theory, she regarded the questions between them as things indifferent. She, the sovereign authority, if the Oxford view of the Church's functions was correct, declared that on such points we might follow our own judgment. This conclusion was forced home upon me, and shook the confidence which I had hitherto continued to feel in Newman. It was much in itself, and it relieved me of other perplexities. The piety, the charity, the moral excellence in the circle into which I had been thrown were evidences as clear as any evidence could be of a living faith. If the Catholic revivalists were right, these graces were but natural virtues, not derived through any recognized channel, uncovenanted mercies, perhaps counterfeits, not virtues at all, but cunning inventions of the adversary. And it had been impossible for me to believe this. A false diamond may gain credit with eyes that have never looked upon the genuine gem, but the pure water once seen cannot be mistaken. More beautiful human characters than those of my Irish Evangelical friends I had never seen, and I have never seen since. Whatever might be the 'Notes of the Church,' a holy life was the first and last of them; and a holy life it was demonstratedly plain to me, was no monopoly of the sacramental system.

At the end of a year[5] I returned to Oxford. There had been a hurricane in the interval, and the storm was still raging. Not the University only, but all England, lay and clerical, was agitating itself over Tract XC. The Anglican Church had been long ago described as having a Catholic Prayerbook, an Arminian clergy, and Calvinistic Articles. When either of the three schools asserted itself with emphasis the others took alarm. Since the revolution of 1688 Church and clergy had been contented to acquiesce in the common title of Protestant; by consent of high and low the very name of Catholic had been abandoned to the Romanists; and now when a Catholic party had risen again, declaring that they and they

[5] [Tract XC was published in 1841.]

only were true Church of England men, the Articles,[6] not unnaturally, had been thrown in their teeth. All the clergy had subscribed the Articles. The Articles certainly on the face of them condemned the doctrines which the revivalists had been putting forward. Weak brothers among them were beginning to think that the Articles had committed the Church to heresy, and that they ought to secede. There were even a few who considered that their position was not so much as honest. I recollect the professor of Astronomy saying to me about this time that the obligation of a Tractarian to go to Rome was in the ratio of his intellectual obtuseness. If he was clever enough to believe two contradictory propositions at the same time, he might stay in the Church of England; if his capacity of reconciliation was limited, he ought to leave it. It was to soothe the consciousness of these troubled spirits that Tract XC was written. As their minds had opened they had recognized in the mass, in purgatory, in authority of tradition, in infallibility of councils, doctrines which down to the schism had been the ancient faith of Christendom. The Articles seemed distinctly to repudiate them; and if these doctrines were true the body which rejected them could be no authentic branch of the Church Catholic. Newman undertook to remove this difficulty. He set himself to ' minimize ' what the Articles said, just as in latter years he has ' minimized ' the decree of Papal infallibility. He tells us that he cannot understand a religion which is not dogmatic; but he too finds tight-lacing uncomfortable; and though he cannot do without his dogma, it must mean as little as possible for him. He argues, in the first place, that the Articles could not have been intended to contradict the canons of the Council of Trent,[7] as was popularly supposed, because they had been composed several years before those canons were published or the Council itself completed. Secondly, that they were directed not against Catholic doctrines, but against the popular abuses of those doctrines. They condemned ' masses;' they did not condemn the mass. They condemned

---

[6] [Embodied in the Elizabethan Act of Supremacy (1559) and enforced after 1571.]

[7] [1545-1563.]

the Romish doctrine of purgatory; but the Romish was not the Greek, and there might be many others. Finally, the Articles were legal documents, and were to be interpreted according to the strict meaning of the words. We do not interpret an Act of Parliament by what we know from other sources of the opinions of its framers; we keep to the four corners of the Act itself. Newman said that we had as little occasion to trouble ourselves with the views of individual bishops in the sixteenth century.

The English mind does not like evasion; and on its first appearance the Tract was universally condemned as dishonest. Very good people, my Irish friends among them, detested it, not for the views which it advocated, but as trifling with truth. I could not go along with them, partly because it had become plain to me that, little as they knew it, they themselves had at least equally to strain the language of the Baptismal Service, and of one of the three absolutions; partly because I considered Newman's arguments to be legally sound. Formulas agreed on in councils and committees are not the produce of any one mind or of any one party. They are compromises in which opposing schools of thought are brought at last to agree after many discussions and alterations. Expressions intended to be plain and emphatic, are qualified to satisfy objectors. The emphasis of phrases may remain, but the point emphasized has been blunted. The closer all such documents are scrutinized the more clear becomes the nature of their origin. Certainly, if the Catholic theory is correct, and if the Holy Spirit really instructs mankind through the medium of councils, and therefore decrees which have been shaped in a manner so human, one can but wonder at the method that has been chosen. It seems like a deliberate contrivance to say nothing in seeming to say much; for there are few forms of words which cannot be perforated by an acute legal intellect. But as far as Tract XC was concerned, public opinion, after taking time to reflect, has pronounced Newman acquitted. It is historically certain that Elizabeth and her ministers intentionally framed the Church formulas so as to enable every one to use them who would disclaim allegiance to the Pope. The English Catholics, who were then more than half the nation, applied to the Council of Trent for leave to attend the English Church services, on the express ground that no Catholic doctrine was denied in them. The Council of Trent refused permission, and the petitioners, after hesitating till, in the defeat of the Armada Providence had declared for the Queen, conformed (the greater number of them) on their own terms. They had fought for the Crown in the civil wars; they had been defeated, and since the Revolution had no longer existed as a theological party. But Newman was only claiming a position for himself and his friends which had been purposely left open when the constitution of the Anglican Church was formed.

But religious men do not argue like lawyers. The Church of England might have been made intentionally comprehensive three centuries ago, but ever since 1688 it had banished Popery and Popish doctrines. When the Catholics were numerous and dangerous, it might have been prudent to conciliate them; but the battle had been fought out since, and a century and a half of struggles and conspiracies and revolutions and dethroned dynasties were not to go for nothing. Compromise might have dictated the letter of the Articles, but unbroken usage for a hundred and fifty years had created a Protestant interpretation of them which had become itself authoritative. Our fathers had risked their lives to get rid of Romanism. It was not to be allowed to steal into the midst of us again under false colours. So angry men said at the time, and so they acted.

Newman, however, had done his work. He had broken the back of the Articles. He had given the Church of our fathers a shock from which it was not to recover in its old form. He had written his Tract, that he might see whether the Church of England would tolerate Catholic doctrine. Had he waited a few years, till the seed which he had sown could grow, he would have seen the Church unprotestantizing itself more ardently than his most sanguine hope could have anticipated, the squire parsons of the Establishment gone like a dream, an order of priests in their places, with an undress uniform in the world, and at their altars ' celebrating ' masses in symbolic robes, with a directory to guide their inexperience. He would have seen them hearing confession, giving absolution, adoring

Our Lady and professing to receive visits from her, preaching transubstantiation and purgatory and penance and everything which his Tract had claimed for them; founding monasteries and religious orders, washing out of their naves and chancels the last traces of Puritan sacrilege; doing all this in defiance of courts of law and Parliaments and bishops, and forcing the authorities to admit that they cannot be interfered with. It has been a great achievement for a single man; not the less so that, although he admitted that he had no right to leave the Church in which he was born unless she repudiated what he considered to be true, he himself would not even pause to discern whether she would repudiate it or not.

But Newman, though he forbids private judgment to others, seems throughout to retain the right of it for his own guidance. He regarded the immediate treatment of the message which he had delivered as the measure of his own duty. His convictions had grown slowly on himself; they were new to the clergy, unpalatable to the laity, violently at variance with the national feelings and traditions. Yet the bishops were expected to submit on the spot, without objection or hesitation, to the dictation of a single person; and because they spoke with natural alarm and anxiety, his misgivings about the Catholicity of the Church of England turned instantly into certainties, and in four years carried him away over the border to Popery.

It is evident now, on reading Newman's own history of his religious opinions, that the world, which said from the beginning that he was going to Rome, understood him better than he then understood himself, or, perhaps, than he understands himself now. A man of so much ability would never have rushed to conclusions so precipitately merely on account of a few bishops' charges. Excuses these charges might be, or explanations to account for what he was doing, but the motive force which was driving him forward was the overmastering ' idea ' to which he had surrendered himself. He could have seen, if he had pleased, the green blade of the Catholic harvest springing in a thousand fields; at present there is scarcely a clergyman in the country who does not carry upon him in one form or other the marks of the Tractarian movement. The answer which he required has been given. The

Church of England has not only admitted Catholic doctrine, but has rushed into it with extraordinary enthusiasm. He might be expected to have recognized that his impatient departure has been condemned by his own arguments. Yet the ' Apologia ' shows no repentance nor explains the absence of it. He tells us that he has found peace in the Church of Rome, and wonders that he could ever have hoped to find it in the English Communion. Very likely. Others knew how it would be from the first. He did not know it; but if the bench of bishops had been as mild and enduring as their present successors, it would have made no difference.

Newman was living at Littlemore, a village three miles from Oxford, when I came back from Ireland. He had given up his benefice, though still occasionally preaching in St Mary's pulpit before the University. He was otherwise silent and passive, though his retirement was suspected, and he was an object of much impertinent curiosity. For myself he was as fascinating as ever. I still looked on him—I do at this moment—as one of the two most remarkable men whom I have ever met with; but I had learnt from my evangelical experiences that equally good men could take different views in theology, and Newmanism had ceased to have exclusive interest to me. I was beginning to think that it would be well if some of my High Church friends could remember also that opinions were not everything. Many of them were tutors, and tutors responsible for the administration of the University. The discipline was lax, the undergraduates were idle and extravagant : there were scandalous abuses in college management, and life at the University was twice as expensive as it need have been. Here were plain duties lying neglected and unthought of, or, if remembered at all, remembered only by the Liberals, whom Newman so much detested. Intellectually, the controversies to which I had listened had unsettled me. Difficulties had been suggested which I need not have heard of, but out of which some road or other had now to be looked for. I was thrown on my own resources, and began to read hard in modern history and literature. Carlyle's books came across me; by Carlyle I was led to Goethe. I discovered Lessing for myself, and then Neander and Schleiermacher. The ' Vestiges of the Natural History of Creation,' which came out about that time, introduced modern science to us under an unexpected aspect, and opened new avenues of thought. As I had perceived before that the Evangelicals could be as saint-like as Catholics, so now I found that men of the highest gifts and unimpeached purity of life could differ from both by whole diameters in the interpretation of the same phenomena. Further, this became clear to me, that the Catholic revival in Oxford, spontaneous as it seemed, was part of a general movement which was going on all over Europe. In France, in Holland, in Germany, intellect and learning had come to conclusions from which religion and conscience were recoiling. Pious Protestants had trusted themselves upon the Bible as their sole foundation. They found their philosophers and professors assuming that the Bible was a human composition —parts of it of doubtful authenticity, other parts bearing marks on them of the mistaken opinions of the age when these books were written; and they were flying terrified back into the Church from which they had escaped at the Reformation, like ostriches hiding their heads in a bush.

Yet how could the Church, as they called it, save them? If what the philosophers were saying was untrue, it could be met by argument. If the danger was real, they were like men caught in a thunder-storm, flying for refuge to a tree, which only the more certainly would attract the lightning. Catholics are responsible for everything for which Protestants are responsible, plus a great deal besides which Protestants rejected once as lies, and the stroke will fall where the evidence is weakest. Christianity, Catholic and Protestant alike, rests on the credibility of the Gospel history   Verbal inaccuracies, if such there be, no more disprove the principal facts related to the Gospels than mistakes in Lord Clarendon's History of the Rebellion prove that there was never a commonwealth in England. After all is said, these facts must be tested by testimony, like all other facts. The personal experiences of individuals may satisfy themselves, but are no evidence to others. Far less can the Church add to the proof, for the Church rests on the history, not the history on the Church. That the Church exists, and has existed, proves no more than that it is an institution which has had a beginning in time, and may have an end in time. The individuals of whom it is composed have believed in Christianity, and their witness is valuable according to their opportunities, like that of other men, but this is all. That the Church as a body is immortal, and has infallible authority antecedent to proof, is a mere assumption, like the tortoise in the Indian myth. If the facts cannot be established, the Catholic theory falls with the Protestant; if they can, they are the common property of mankind, and to pile upon them the mountains of incredibilities for which the Catholic Church has made itself answerable, is only to play into the hands of unbelievers, and reduce both alike to legend.

Still, the reaction was a fact, visible everywhere, especially in Protestant countries. The bloody stains on the Catholic escutcheon were being painted over. The savage massacres, the stake at Smithfield, and the Spanish auto-da-fé, the assassinations and civil wars and conspiracies at which we had shuddered as children, were being condoned or explained away. Hitherto it had been strenuously denied that the Oxford movement was in the direction of Rome; it was insisted rather that, more than anything else, Tractarianism would tend to keep men away from Rome. No Protestant had spoken harder things of the Roman see and its doings than Newman had and I was still for myself unable to believe that he was on his way to it. But the strongest swimmers who are in the current of a stream must go where it carries them, and his retirement from active service in the Church of England showed that he himself was no longer confident.

ON THE USES OF A LANDED GENTRY[1]

Before I proceed with the address which I have undertaken to deliver this evening, I ought to explain why I have chosen a subject which lies outside the usual lines. Your institution is philosophical not political, and these lectures are properly confined to subjects on which, if we cannot agree, we ought to be able to agree to differ.

I might say that the question of the Uses to the Community of a Landed Gentry is in this country purely a philosophical one. It is certainly not a practical question. It is no question of practical politics. No reasonable man that I know of seriously wishes for an agrarian law, or for a forcible division of landed property, or for an interference with the right of making settlements, or with our right to make our own wills in whatever way may seem good to us.

There are persons, perhaps, who do not like the way in which the land is apportioned—who would wish it was more evenly shared among the people. But they wish it only as they might wish that we had a drier or a milder climate. There are others who are well satisfied with things as they are; who have no objection to the large estates, who do not quarrel with primogeniture, who are well satisfied with entails, particularly if they have the happiness of benefiting by them. But they do not like the subject to be talked about, and would prefer that it should be judiciously let alone. I cannot see that there is any need for reticence. In a free country like ours the distribution of land depends on economic laws as absolute as the law of gravity. So long as the British nation continues as it is, the landed gentry are as fixed a part of it as the planets of the solar system. Individuals may fall from their spheres and ruin themselves by their own folly. The institution itself is as secure as the succession of seasons as long as the inclination of the pole remains unaltered.

[1] Address delivered at the Philosophical Institute at Edinburgh, November 6, 1876.

Why this should be is an interesting problem of social philosophy well deserving more enquiry than it has received. My own object, however, when I originally thought of addressing you on the landed gentry, was far less ambitious. Three years ago, when your directors were kind enough to ask me to come here, the misdoings of a certain class of landlords had been much talked about in connection with the Irish Land Act. An acquaintance of my own, Mr Smith, of the Scilly Isles, had recently died. Mr Smith had possessed exceptional and unusual powers in those islands, and had not abused them. I thought that at such a time an account of such a man and his doings might not be unwelcome as an evidence that a landlord was not necessarily as pernicious a being as some people appeared to think. It used to be said before the American war that masters who were kind to their slaves were the worst enemies that the slaves had. They had made the apology for a detestable institution. I have heard the same objection made of good husbands by advanced advocates of the rights of women. On similar grounds a bad opinion may be formed of Mr Smith, and now at this distance of time I do not mean to trouble you at any length about him. Circumstances prevented my taking up the subject when it would have been more to the purpose, and the years which have since rolled by have brought other interests with them. Good actions do to some extent serve as salt to keep a man's memory fresh, but the world is for the living and not for the dead. A very few words on Mr Smith are all to which I shall ask you to listen.

I will bespeak your good opinion for him by saying first that he was an advanced Radical. He was a believer in Bentham. 'The greatest happiness of the greatest number' was the rule of his life. Besides his property in Scilly he had an estate at Berkhampstead, in Hertfordshire. At Berk-hamstead there is an extensive common, one of the few great commons remaining in England, a free expanse of grass and forest, much valued by the country-side and by all the neighbourhood. On either side of it lay the estates of a great nobleman, inconveniently divided by the intruding wilderness; and the inconvenience of the noble lord himself assumed the man turneth away,' &c. The clerk started up in his seat and said, 'I beg your pardon, sir, he is not come yet.'

This was the rule in Scilly when I was there. The Lord of the Isles, as Mr Smith was called, was supreme in Church as well as State. He is gone now. Another king rules in his stead. I trust he may prove a wicked man too, like his uncle.

This may be all very well, says my Radical friend, but we cannot keep up a system which gives one man a power over the fortunes of thousands because one in a hundred may now and then make a wholesome use of it. It might answer when the nation was half-grown. We are of age now, and have done with leading-strings. The land belongs to the people. No limited number of persons have a right to raise fences round their thousands or ten thousands of acres, and say, 'This land is mine. None but I shall enter upon it.' The soil is the common inheritance of all sons of Adam who are born into the world. The way to improve landlords is to improve them out of existence.

The same idea was once expressed to me by Mr Hartley Coleridge. 'Property!' he said, 'I hate the word; because I have not got any of my own.'

Of course every one born into this world must live on the land, and be fed on what the land produces; at least outside China, where a few millions, I believe, live in barges and are fed on fish. But we don't want a general scramble. There must be some arrangement. The Socialist says the land should be held by the State, and be portioned out to those who will cultivate it. Is the State to resume these portions at its pleasure? If yes, what becomes of personal liberty? If no, you have a multitude of small proprietors instead of a few large ones. And what is to prevent them from selling their interest, and the large estates from growing again? In Great ·Britain and among the British people such as we know them, you may divide the land as you please; but if you leave personal liberty the phenomena which you deprecate are certain to recur.

A few years ago there was a loud outcry at what was called the monopoly of land. Twelve noblemen were said to own half Scotland, a few hundreds to own half England. The

aspect of a public injury. A common which belonged to the people, appeared to him to belong to no one in particular. He meant no harm. He was incapable of doing anything which he did not believe to be just; but he was informed by those who managed his estates for him, that it would be to the general advantage if the occasion of so much disorder was taken away. He doubted the result of an appeal to law, but a plea was found which he hoped might sustain him if he was once in possession. He fenced the common in, and he left the people of Berkhampstead to find their remedy. The smaller landowners, as he expected, did not like to quarrel with their powerful neighbour. The poor, who were the most injured, had the least means of protecting themselves, and Berkhampstead Common would have gone the way of a hundred others except for Mr Augustus Smith. Mr Smith heard what had been done. He perceived that the advantage would be with the party which was actually in occupation. Instead of bringing an action against the noble lord, he brought a hundred and fifty navvies one dark night down from London. When morning came fifteen hundred yards of iron railing were lying flat upon the ground. They were never set up again, and Berkhampstead Common still belongs to you and to me, and to any one who chooses to enjoy himself there.

But now for what Mr Smith did in Scilly. The Scilly Isles are a prolongation of the granite backbone of Devonshire and Cornwall, and are, in fact, but a cluster of granite hill-tops standing out of the water. The largest island is from four to six miles round. Three others are about half that size; the rest, some hundreds in number, are little more than rocks. Before the Reformation, Scilly was occupied by monks, who had a fancy for such places. When the monks went it became a pirates' nest, and then a haunt of privateers and smugglers. After the great war it sank into the condition of some of your own western islands. The population was large, as it always is where there is no motive for prudence. The people were miserably poor; they lived in squalid hovels, with a half acre or an acre of ground, which they manured with seaweed. They eked out their livelihood by fishing, piloting, and occasional smuggling ventures. They had no schools, and they had public-houses; and spirits were cheap where customs duties were so easily evaded.

The Crown is the owner of these islands. Circumstances, about forty years ago, induced Mr Smith to take a long lease of them. As sole lessee he became absolute master there; and if any one wishes to see what can be done by one man of no extraordinary abilities, but with a strong will and a resolute purpose to do good, let him employ his next summer holiday in paying Scilly a visit.

Mr Smith at once altered the small tenures so as to make improvement possible. He broke up the small holdings and combined them into farms on which a family could be maintained in decency. He provided work at competent wages for those who were deprived of their potato patches. He drained. He enclosed the fields. He rebuilt the cottages in a form fit for human beings. He set up boat-yards, and organised the fishing business. He stopped drunkenness with a high hand. Incorrigible blackguards he shipped off to the mainland. He built chapels and endowed them. He built schools and provided proper teachers for them. The young lads were trained generally for the sea, and with such effect that when I last enquired I was told that the Scilly pilots had the best name of all the pilots at the mouth of the Channel, and that there was not a Scilly boy in the merchant service, above twenty-one, who was a sailor before the mast; all were masters or petty officers.

The soil, properly cultivated, began to produce unheard-of crops. The soft, warm climate brings vegetation forward early, and the Scilly gardeners are now making their fortunes by supplying spring vegetables to the London market. Throughout the compass of the British Islands you will not find an equal number of people on an equal area, on an average, so well clothed, so well fed, so well lodged, so well educated. In the largest island there is but one constable, and he is the only person there who has nothing to do. The whole place wears—or did wear when I was there—an air of quiet industry, prosperity, order, and discipline.

These results Mr Smith arrived at by the arbitrary exercise of his power as landlord. He was a Radical who looked to

ends rather than means. He desired to promote the great happiness of the people dependent on him, and he took the readiest road to his object. He found Scilly a rabbit warren of paupers. He made it a thriving community of industrious men and women. If boys and girls wanted to marry, and could not show that they were in a condition to support a family, he told them that he had no room for them; they must wait till they had money at the savings' bank, or they must move off to the mainland. He was a king on a small scale. Within the law his authority was absolute, and he used it not for himself but for his subjects. He made no money in Scilly. He told me a few years before his death that he had laid out more there than he had ever received. He was a thrifty man in his own habits, and had few luxuries but his garden. His rents he spent upon the people, and when he died he left the islands trebled in mere money value.

'There is prosperity of a kind, undoubtedly,' said a philosophic Radical to me, who had been to Scilly to study what was going on; 'but it is paternal government. I detest paternal government.' Paternal government may be detestable where you have the wrong sort of father. Men like Mr Smith are rare; but I am none the less thankful when a rare chance gives the right man the right opportunity. If the islanders had been as free as Mr Mill would have desired to see them, and if they had been all animated with the most determined spirit of self-improvement, they could not have accomplished in a hundred years what Mr Smith accomplished for them in one generation. He valued liberty as much as any man when liberty meant resistance to what was wrong. He was less patient of liberty to resist what was clearly and indisputably right.

He had his foibles. He was the wicked man of the islands. You know the story of the wicked man. It is so old that perhaps I ought not to mention it. A clergyman of the Church of England had taken a friend's duty in a parish where there was a despotic squire who did not allow the service to be commenced before his arrival. The clergyman, not knowing the custom, began at the proper hour with the opening words of the English Liturgy, 'When the wicked

quarter of a million freeholders who existed in Queen Anne's time were supposed to have dwindled to thirty thousand, and their numbers to be yearly diminishing. An enquiry was made. We have a new Domesday Book, and it appears that instead of no more than thirty thousand freeholders in this country, we have nearly a million. Yet the details, when looked into, do in part bear out what the agitators complained of. The House of Lords does own more than a third of the whole area of Great Britain. Two-thirds of it really belongs to great peers and commoners, whose estates are continually devouring the small estates adjoining them. The remaining third, in and about the great towns, is subdivided, and the subdivision is continually increasing, but the land there also is still falling mainly into the hands of the rich.

Near the cities spade cultivation answers from the ready market for garden produce; and small free-holds, purely agricultural, are held in this way. But in general rich speculators buy land about the cities for building, and bid high for it. Successful tradesmen, merchants, or manufacturers want houses of their own in the neighbourhood, outside the smoke, with gardens and a small dairy-farm as a luxury. Under these conditions the small holdings multiply.

At a distance from the cities we have exactly the opposite. Agricultural land, on an average, pays but two per cent. interest on its selling value. A yeoman cultivating his own land finds it to his advantage to sell it, rent it from some one else, and employ his purchase money in his business. A young Scotch or Englishman, coming into possession of an estate worth a few hundreds a year, if he has any spirit in him, does not settle down upon it in obscurity. He sells his scanty acres, takes his capital with him, and invests it where he can get some better return, or he goes into trade or emigrates. There are these two tendencies in operation which you cannot interfere with while you leave us our liberty, and both of them give the land to those who can afford to pay for it as a luxury.

Will you tell the embarrassed owner of a small property that he is not to sell it? The law of entail does say this in some instances, and so tends to preserve the small properties. People complain of the law of entail as if it interfered with the subdivision of landed property. It rather sustains such small estates as remain. Abolish entail if you please, but accumulation will only proceed the more rapidly.

Will you tell a large landowner that he is not to buy a property adjoining his own, when he will give a higher price for it than any one else? You cannot do this without robbing the person who wishes to sell.

Will you have the Code Napoléon? Will you insist that when a landowner dies his estate shall be divided among his children? If you were to pass such a law you would fail still to produce the effect which is produced in France, because the British and French people are essentially different. The home of the French peasant is France, and he will thrive nowhere else. The home of the Scot or the Englishman is the whole globe. Three centuries ago we were confined within our own four seas. Where are we now? We have spread over North America. We are filling Australia, New Zealand, South Africa. There is scarcely a seaport in either hemisphere where you will not find an English-speaking community. I once heard a discussion at a *table-d'hôte* at Madrid, between twenty or thirty commercial travellers, as to which language was of most use to them. There was not an Englishman in the party, but they all agreed that the English language would carry them farthest. Create your small landed proprietors by law, and the energetic among them will still sell, and carry their capital to a better market.

Primogeniture! you will say. At least there ought to be no primogeniture. Why make a distinction between personal property and real property? Why should an eldest son be preferred to his brothers and sisters, to his own injury and theirs? Abolish primogeniture by all means if you can. I need not say a word in favour of it, but understand what it is that you would change. It is not a law. It is a custom. The law gives the land to the eldest son if the father dies without a will. But he need not die without a will. He can divide his land among his children if he pleases. The fact is that he does not please. Primogeniture is the custom which he follows and assists to make. The law does not take effect in one case out of a thousand. Even the law is not universally the same. The Saxon gavelkind[2] remains in Kent. But the practice in Kent is the same as the practice elsewhere. Men leave their lands to their eldest sons because they wish to preserve their families. If you want a change you must alter their nature, or else you must take away their liberty.

Again, it is said the conveyance of land ought to be easier than it is. In other countries you can buy a piece of land as easily as a yard of calico. In England the process is so expensive as to put a few acres beyond a poor man's reach. You may cheapen conveyancing, yet the poor man will still not get his acres. The more easy the transfer, the faster the land will flow in the channels which it tends of itself to follow.

But the less obstruction the better. Let us have free trade in land by all means, as in everything else. There is but one serious objection that I know of. I cannot tell how it may suit the lawyers. When the Reformation began in England, the House of Commons complained to the Crown of the enormous expenses of the Ecclesiastical Courts. The Archbishop of Canterbury said in reply, that no doubt the proceedings in the courts were costly, but the costs went to maintain a very excellent class of persons, without whom the country would be exceedingly ill off, the learned gentlemen of the long robe. There is force in this answer. I should be sorry to say anything against it. One of the most valuable lessons which I have learnt in life is the prudence of keeping on good terms with the lawyers.

On the whole it seems to me certain that unless the area of Great Britain could be made larger than it is, or until the British people change their nature, a peasant proprietary is a dream. So long as a free energetic race of men are crowded together in a small space with every variety of employment open to them at home, with wide avenues to distinction offering themselves abroad, and with every individual striving to push his way to a higher station than that in which he was born, so long the ownership of land will be the luxury of the comparatively few. A time I suppose will arrive when the

[2] [The practice of equal division among all the sons.]

giddy whirl of industry and progress will cease among us, when we shall no longer struggle for a first place among the nations. Then the tide will ebb; then the great estates will dissolve, and the soil will again be divided among unambitious agricultural freeholders. The land then will suffice for the support of all who live upon it. The grass will grow in the streets of Manchester. The Clyde will eddy round the rotting wrecks of the Glasgow merchant-ships, and the plough will pass over the gardens of its merchant princes. The reign of Saturn will come back, and the golden age of pastoral simplicity. Till that time comes you must lay your account for a landed gentry of some kind, and accepting the inevitable fact, you must try to make the best of it.

Nor do I think the prospect need much disturb us. Our landed system is like our political system : it consists of a number of petty monarchies, which are gradually becoming restricted by custom, till the monarch shall remain powerful for good and comparatively powerless to hurt. Let us put the worst side of it first. The restraints upon a landlord's power which are not self-imposed by the grant of leases, are still mainly restraints of usage and public opinion, and men are unequally amenable to these influences.

The possession of a large estate carries with it authority which can still be abused, and this authority may fall by the accident of birth to a person unfit to be trusted with it. The young heir is a fool or a spendthrift, and tenants, labourers, every one dependent upon him, suffer in consequence.

Nature provides a remedy of a kind. Folly brings difficulties, and difficulties bankruptcy. The incompetent owner is sold up. Nature shakes him off, and puts a better in his place. Society, like each of ourselves, is perpetually renovating itself. The used-up tissue of our bodies passes away at every moment— young and healthy tissue is growing instead of it. Watch the land tenure in any busy county of England, and you will be surprised to see how rapidly a similar process is going on. I was standing a few years ago on a hill about fifteen miles from London, looking round over the richly cultivated country —dark woods marking here and there the parks and pleasure-grounds of the lords of the soil. I asked my companion, who himself was one of them, how long on an average an estate remained about there in the same family. He answered, perhaps twenty years.

Again, there is the wilder remedy which we used to hear of in the sister island. The landlord may become a direct oppressor. He may care nothing for the people, and have no object but to squeeze the most that he can out of them, fairly or unfairly. The Russian Government has been called despotism tempered by assassination. In Ireland for many years landlordism was tempered by assassination.

Unfortunately the wrong man was generally assassinated. The true criminal was an absentee, and his agent was shot instead of him. A noble lord living in England, two of whose agents had lost their lives already in his service, ordered the next to post a notice in his Barony that he intended to persevere in what he was doing, and if the tenants thought they would intimidate him by shooting his agents, they would find themselves mistaken.

Thus the desired result was not effected, and Ireland could not be left to natural remedies; every circumstance combined in that country to exasperate the relations between landlord and tenant. The landlords were, for the most part, aliens in blood and aliens in religion. They represented conquest and confiscation, and they had gone on from generation to generation with an indifference to the welfare of the people which would not have been tolerated in England and Scotland. The law had to interfere at last to protect the peasantry in the shape of Mr Gladstone's Land Act; the best measure, perhaps the only good measure, which has been passed for Ireland for the last two hundred years.

In Ireland there are good landlords, more than are ever heard of. The object of Mr Gladstone's Act[3] was simply to shape a law out of the good landlords' practice and make the bad conform to it. It was called confiscation; I know not what was confiscated. Nothing certainly to which the landlord had any equitable right. The selling value of land has not

[3] [1870. This made eviction illegal except for non-payment of rent; it also provided compensation for improvements and loans for purchase of holdings.]

diminished in Ireland since that Act was passed. It has rather risen from the increase of security.

It is possible that a similar law may become necessary in England and Scotland. It is possible, if infinitely improbable. Responsibility is the shadow of a great position. If a time should ever come when the heirs of great estates forget that they *have* any responsibility, if they come to suppose that the world was made for them, not they for the world, that the sole duty laid on them is the duty of enjoying themselves, that they are permitted to idle away a life made weary to them by its inanity between the London season, the foreign watering-place, the deer forest, the battue, or a salmon river; then it is easy to prove that an end will come to all that.

I was staying the year before the Irish famine at a large house in Connaught. We had a great gathering there of the gentlemen of the county; more than a hundred of us sat down to a luncheon on the lawn. My neighbour at the table was a Scotchman, who was over there examining the capabilities of the soil. ' There,' he said to me, ' you see the landed gentry of this county. In all the number there may be one, at the most two, who believe that the Almighty put them into this world for any purpose but to shoot grouse, race, gamble, drink, or break their necks in the hunting-field. They are not here at all for such purposes, and one day they will find it so.'

The day of reckoning was nearer than he thought. Next year came the potato disease. The estates of most of them were mortgaged, and at best they had only a margin to live upon. Rents could not be paid. The poor people were dying of hunger, and a poor-rate had to be laid on amounting, in places, to confiscation. The Encumbered Estates Act followed, and the whole set of them were swept clean away.

We are not come to that pass here, nor do I believe we are likely to come. Even here we have heard occasionally of strange things being done; uncalled-for evictions of tenantry, with mountain and glen closed against the tourist and artist, that a noble lord and his friends may shoot a few miserable deer. But the tendency of things is not to an increase of all that, very far otherwise.

Another noble lord that I know of has a mountain property in Kerry which would make the finest deer-forest in these islands. He has the same temptation to make a deer-forest as those others have. As a forest it would bring him five times the present rent. Some forty or fifty families only would have to be removed from their farms, and they could be bought out under the Land Act with enormous profit to the landlord. But the deer are not on the mountains, and those families remain on their farms. The same noble lord spends four-fifths of the income which he draws from that property in improving the condition of the people. No one speaks of this; no one ever talks of what is done wisely and well. Health is never conscious of itself. We are only conscious of our own bodies when something is amiss with us. Offences only attract notice. Judge of British society from the police reports, and we are a nation of savages. Yet we are always forgetting this. We hear of the bad exceptions. We hear of them because they are exceptions, and we argue as if they were the rule.

Well then, gentlemen, let us turn from the mischief which may come of a landed gentry, and let us see what good comes of them.

Since land does not pay as a commercial speculation, why do rich men give such large prices for it? Land is sought after for the social consequence and for the political influence which the possession of a large estate in such a country as ours confers. It is sought after from an ambition to leave our names behind us, rooted into the soil to which the national life is attached. To obtain or keep such a position, money must be sacrificed to other considerations; and the sacrifice must be maintained and continued if the landowner is to preserve the objects for which it is made. The same force in nature appears now as heat, now as motion; one can be converted into the other. Wealth in the same way may appear in the form of luxury, or it may appear in the form of power. The landowner who desires honour and influence spends the rents which fall to him rather as a revenue than as a private income. The manager of the estates of a noble duke who is nominally one of the richest men in Great Britain said to me, that in his experience dukes never had any money.

On those estates more than a million had been laid out in a few years in rebuilding the cottages.

And the farther what is called the land monopoly is carried, the more, that is, the small estates are absorbed in the large, the better these duties will be performed. I don't know how it may be in Scotland, but I know that in England you can tell by the look of the country which you are passing through whether it belongs to a large landowner or a small one.

Compare an estate owned by one man with a hundred thousand a year, and a similar estate divided among a hundred owners with a thousand a year each. On which of these will the working tenants find themselves best off? The one great man's establishment may be expensive, but after all it is but one. The expenses of the most splendid household will not reach a hundred thousand a year, or half that sum, or a quarter of it. The great man is on a pedestal. If he is evil spoken of his pedestal becomes a pillory. Therefore he does not press his rights when he might press them. The customs of the manor are generally observed. Farm buildings are kept in good condition, fences are in good repair, cottages have roofs which will keep the rain out. You find churches, you find schools, you find everything which public opinion demands or approves.

Turn to the estate which is divided between the hundred less conspicuous proprietors. Will an equal margin of income be forthcoming for improvements? Will there be the same consideration for tenants and labourers? There cannot be, because a hundred private establishments have to be supported instead of one, and a hundred families struggling to maintain the position of gentry with inadequate means. By them every farthing which their estate will yield is required for their ordinary expenditure. They are embarrassed. They must borrow. Their obvious duties are left undone. You read the story in unmended fences, in broken gates, in decaying farm-houses. At length a crisis comes, and unless entail interferes the land is sold to some one who can better afford to keep it.

Latifundia perdidere Italiam—the great estates ruined Italy.

had public-houses; and spirits were cheap where customs duties were so easily evaded.

The Crown is the owner of these islands. Circumstances, about forty years ago, induced Mr Smith to take a long lease of them. As sole lessee he became absolute master there; and if any one wishes to see what can be done by one man of no extraordinary abilities, but with a strong will and a resolute purpose to do good, let him employ his next summer holiday in paying Scilly a visit.

Mr Smith at once altered the small tenures so as to make improvement possible. He broke up the small holdings and combined them into farms on which a family could be maintained in decency. He provided work at competent wages for those who were deprived of their potato patches. He drained. He enclosed the fields. He rebuilt the cottages in a form fit for human beings. He set up boat-yards, and organised the fishing business. He stopped drunkenness with a high hand. Incorrigible blackguards he shipped off to the mainland. He built chapels and endowed them. He built schools and provided proper teachers for them. The young lads were trained generally for the sea, and with such effect that when I last enquired I was told that the Scilly pilots had the best name of all the pilots at the mouth of the Channel, and that there was not a Scilly boy in the merchant service, above twenty-one, who was a sailor before the mast; all were masters or petty officers.

The soil, properly cultivated, began to produce unheard-of crops. The soft, warm climate brings vegetation forward early, and the Scilly gardeners are now making their fortunes by supplying spring vegetables to the London market. Throughout the compass of the British Islands you will not find an equal number of people on an equal area, on an average, so well clothed, so well fed, so well lodged, so well educated. In the largest island there is but one constable, and he is the only person there who has nothing to do. The whole place wears—or did wear when I was there—an air of quiet industry, prosperity, order, and discipline.

These results Mr Smith arrived at by the arbitrary exercise of his power as landlord. He was a Radical who looked to

aspect of a public injury. A common which belonged to the people, appeared to him to belong to no one in particular. He meant no harm. He was incapable of doing anything which he did not believe to be just; but he was informed by those who managed his estates for him, that it would be to the general advantage if the occasion of so much disorder was taken away. He doubted the result of an appeal to law, but a plea was found which he hoped might sustain him if he was once in possession. He fenced the common in, and he left the people of Berkhampstead to find their remedy. The smaller landowners, as he expected, did not like to quarrel with their powerful neighbour. The poor, who were the most injured, had the least means of protecting themselves, and Berkhampstead Common would have gone the way of a hundred others except for Mr Augustus Smith. Mr Smith heard what had been done. He perceived that the advantage would be with the party which was actually in occupation. Instead of bringing an action against the noble lord, he brought a hundred and fifty navvies one dark night down from London. When morning came fifteen hundred yards of iron railing were lying flat upon the ground. They were never set up again, and Berkhampstead Common still belongs to you and to me, and to any one who chooses to enjoy himself there.

But now for what Mr Smith did in Scilly. The Scilly Isles are a prolongation of the granite backbone of Devonshire and Cornwall, and are, in fact, but a cluster of granite hill-tops standing out of the water. The largest island is from four to six miles round. Three others are about half that size; the rest, some hundreds in number, are little more than rocks. Before the Reformation, Scilly was occupied by monks, who had a fancy for such places. When the monks went it became a pirates' nest, and then a haunt of privateers and smugglers. After the great war it sank into the condition of some of your own western islands. The population was large, as it always is where there is no motive for prudence. The people were miserably poor; they lived in squalid hovels, with a half acre or an acre of ground, which they manured with seaweed. They eked out their livelihood by fishing, piloting, and occasional smuggling ventures. They had no schools, and they

man turneth away,' &c. The clerk started up in his seat and said, ' I beg your pardon, sir, he is not come yet.'

This was the rule in Scilly when I was there. The Lord of the Isles, as Mr Smith was called, was supreme in Church as well as State. He is gone now. Another king rules in his stead. I trust he may prove a wicked man too, like his uncle.

This may be all very well, says my Radical friend, but we cannot keep up a system which gives one man a power over the fortunes of thousands because one in a hundred may now and then make a wholesome use of it. It might answer when the nation was half-grown. We are of age now, and have done with leading-strings. The land belongs to the people. No limited number of persons have a right to raise fences round their thousands or ten thousands of acres, and say, ' This land is mine. None but I shall enter upon it.' The soil is the common inheritance of all sons of Adam who are born into the world. The way to improve landlords is to improve them out of existence.

The same idea was once expressed to me by Mr Hartley Coleridge. ' Property!' he said, ' I hate the word; because I have not got any of my own.'

Of course every one born into this world must live on the land, and be fed on what the land produces; at least outside China, where a few millions, I believe, live in barges and are fed on fish. But we don't want a general scramble. There must be some arrangement. The Socialist says the land should be held by the State, and be portioned out to those who will cultivate it. Is the State to resume these portions at its pleasure? If yes, what becomes of personal liberty? If no, you have a multitude of small proprietors instead of a few large ones. And what is to prevent them from selling their interest, and the large estates from growing again? In Great ·Britain and among the British people such as we know them, you may divide the land as you please; but if you leave personal liberty the phenomena which you deprecate are certain to recur.

A few years ago there was a loud outcry at what was called the monopoly of land. Twelve noblemen were said to own half Scotland, a few hundreds to own half England. The

ends rather than means. He desired to promote the greatest happiness of the people dependent on him, and he took the readiest road to his object. He found Scilly a rabbit warren of paupers. He made it a thriving community of industrious men and women. If boys and girls wanted to marry, and could not show that they were in a condition to support a family, he told them that he had no room for them; they must wait till they had money at the savings' bank, or they must move off to the mainland. He was a king on a small scale. Within the law his authority was absolute, and he used it not for himself but for his subjects. He made no money in Scilly. He told me a few years before his death that he had laid out more there than he had ever received. He was a thrifty man in his own habits, and had few luxuries but his garden. His rents he spent upon the people, and when he died he left the islands trebled in mere money value.

'There is prosperity of a kind, undoubtedly,' said a philosophic Radical to me, who had been to Scilly to study what was going on; 'but it is paternal government. I detest paternal government.' Paternal government may be detestable where you have the wrong sort of father. Men like Mr Smith are rare; but I am none the less thankful when a rare chance gives the right man the right opportunity. If the islanders had been as free as Mr Mill would have desired to see them, and if they had been all animated with the most determined spirit of self-improvement, they could not have accomplished in a hundred years what Mr Smith accomplished for them in one generation. He valued liberty as much as any man when liberty meant resistance to what was wrong. He was less patient of liberty to resist what was clearly and indisputably right.

He had his foibles. He was the wicked man of the islands. You know the story of the wicked man. It is so old that perhaps I ought not to mention it. A clergyman of the Church of England had taken a friend's duty in a parish where there was a despotic squire who did not allow the service to be commenced before his arrival. The clergyman, not knowing the custom, began at the proper hour with the opening words of the English Liturgy, 'When the wicked

quarter of a million freeholders who existed in Queen Anne's time were supposed to have dwindled to thirty thousand, and their numbers to be yearly diminishing. An enquiry was made. We have a new Domesday Book, and it appears that instead of no more than thirty thousand freeholders in this country, we have nearly a million. Yet the details, when looked into, do in part bear out what the agitators complained of. The House of Lords does own more than a third of the whole area of Great Britain. Two-thirds of it really belongs to great peers and commoners, whose estates are continually devouring the small estates adjoining them. The remaining third, in and about the great towns, is subdivided, and the subdivision is continually increasing, but the land there also is still falling mainly into the hands of the rich.

Near the cities spade cultivation answers from the ready market for garden produce; and small free-holds, purely agricultural, are held in this way. But in general rich speculators buy land about the cities for building, and bid high for it. Successful tradesmen, merchants, or manufacturers want houses of their own in the neighbourhood, outside the smoke, with gardens and a small dairy-farm as a luxury. Under these conditions the small holdings multiply.

At a distance from the cities we have exactly the opposite. Agricultural land, on an average, pays but two per cent. interest on its selling value. A yeoman cultivating his own land finds it to his advantage to sell it, rent it from some one else, and employ his purchase money in his business. A young Scotch or Englishman, coming into possession of an estate worth a few hundreds a year, if he has any spirit in him, does not settle down upon it in obscurity. He sells his scanty acres, takes his capital with him, and invests it where he can get some better return, or he goes into trade or emigrates. There are these two tendencies in operation which you cannot interfere with while you leave us our liberty, and both of them give the land to those who can afford to pay for it as a luxury.

Will you tell the embarrassed owner of a small property that he is not to sell it? The law of entail does say this in some instances, and so tends to preserve the small properties. People

complain of the law of entail as if it interfered with the subdivision of landed property. It rather sustains such small estates as remain. Abolish entail if you please, but accumulation will only proceed the more rapidly.

Will you tell a large landowner that he is not to buy a property adjoining his own, when he will give a higher price for it than any one else? You cannot do this without robbing the person who wishes to sell.

Will you have the Code Napoléon? Will you insist that when a landowner dies his estate shall be divided among his children? If you were to pass such a law you would fail still to produce the effect which is produced in France, because the British and French people are essentially different. The home of the French peasant is France, and he will thrive nowhere else. The home of the Scot or the Englishman is the whole globe. Three centuries ago we were confined within our own four seas. Where are we now? We have spread over North America. We are filling Australia, New Zealand, South Africa. There is scarcely a seaport in either hemisphere where you will not find an English-speaking community. I once heard a discussion at a *table-d'hôte* at Madrid, between twenty or thirty commercial travellers, as to which language was of most use to them. There was not an Englishman in the party, but they all agreed that the English language would carry them farthest. Create your small landed proprietors by law, and the energetic among them will still sell, and carry their capital to a better market.

Primogeniture! you will say. At least there ought to be no primogeniture. Why make a distinction between personal property and real property? Why should an eldest son be preferred to his brothers and sisters, to his own injury and theirs? Abolish primogeniture by all means if you can. I need not say a word in favour of it, but understand what it is that you would change. It is not a law. It is a custom. The law gives the land to the eldest son if the father dies without a will. But he need not die without a will. He can divide his land among his children if he pleases. The fact is that he does not please. Primogeniture is the custom which he follows and assists to make. The law does not take effect in

one case out of a thousand. Even the law is not universally the same. The Saxon gavelkind[2] remains in Kent. But the practice in Kent is the same as the practice elsewhere. Men leave their lands to their eldest sons because they wish to preserve their families. If you want a change you must alter their nature, or else you must take away their liberty.

Again, it is said the conveyance of land ought to be easier than it is. In other countries you can buy a piece of land as easily as a yard of calico. In England the process is so expensive as to put a few acres beyond a poor man's reach. You may cheapen conveyancing, yet the poor man will still not get his acres. The more easy the transfer, the faster the land will flow in the channels which it tends of itself to follow.

But the less obstruction the better. Let us have free trade in land by all means, as in everything else. There is but one serious objection that I know of. I cannot tell how it may suit the lawyers. When the Reformation began in England, the House of Commons complained to the Crown of the enormous expenses of the Ecclesiastical Courts. The Archbishop of Canterbury said in reply, that no doubt the proceedings in the courts were costly, but the costs went to maintain a very excellent class of persons, without whom the country would be exceedingly ill off, the learned gentlemen of the long robe. There is force in this answer. I should be sorry to say anything against it. One of the most valuable lessons which I have learnt in life is the prudence of keeping on good terms with the lawyers.

On the whole it seems to me certain that unless the area of Great Britain could be made larger than it is, or until the British people change their nature, a peasant proprietary is a dream. So long as a free energetic race of men are crowded together in a small space with every variety of employment open to them at home, with wide avenues to distinction offering themselves abroad, and with every individual striving to push his way to a higher station than that in which he was born, so long the ownership of land will be the luxury of the comparatively few. A time I suppose will arrive when the

[2] [The practice of equal division among all the sons.]

giddy whirl of industry and progress will cease among us, when we shall no longer struggle for a first place among the nations. Then the tide will ebb; then the great estates will dissolve, and the soil will again be divided among unambitious agricultural freeholders. The land then will suffice for the support of all who live upon it. The grass will grow in the streets of Manchester. The Clyde will eddy round the rotting wrecks of the Glasgow merchant-ships, and the plough will pass over the gardens of its merchant princes. The reign of Saturn will come back, and the golden age of pastoral simplicity. Till that time comes you must lay your account for a landed gentry of some kind, and accepting the inevitable fact, you must try to make the best of it.

Nor do I think the prospect need much disturb us. Our landed system is like our political system: it consists of a number of petty monarchies, which are gradually becoming restricted by custom, till the monarch shall remain powerful for good and comparatively powerless to hurt. Let us put the worst side of it first. The restraints upon a landlord's power which are not self-imposed by the grant of leases, are still mainly restraints of usage and public opinion, and men are unequally amenable to these influences.

The possession of a large estate carries with it authority which can still be abused, and this authority may fall by the accident of birth to a person unfit to be trusted with it. The young heir is a fool or a spendthrift, and tenants, labourers, every one dependent upon him, suffer in consequence.

Nature provides a remedy of a kind. Folly brings difficulties, and difficulties bankruptcy. The incompetent owner is sold up. Nature shakes him off, and puts a better in his place. Society, like each of ourselves, is perpetually renovating itself. The used-up tissue of our bodies passes away at every moment— young and healthy tissue is growing instead of it. Watch the land tenure in any busy county of England, and you will be surprised to see how rapidly a similar process is going on. I was standing a few years ago on a hill about fifteen miles from London, looking round over the richly cultivated country —dark woods marking here and there the parks and pleasure-grounds of the lords of the soil. I asked my companion, who

himself was one of them, how long on an average an estate remained about there in the same family. He answered, perhaps twenty years.

Again, there is the wilder remedy which we used to hear of in the sister island. The landlord may become a direct oppressor. He may care nothing for the people, and have no object but to squeeze the most that he can out of them, fairly or unfairly. The Russian Government has been called despotism tempered by assassination. In Ireland for many years landlordism was tempered by assassination.

Unfortunately the wrong man was generally assassinated. The true criminal was an absentee, and his agent was shot instead of him. A noble lord living in England, two of whose agents had lost their lives already in his service, ordered the next to post a notice in his Barony that he intended to persevere in what he was doing, and if the tenants thought they would intimidate him by shooting his agents, they would find themselves mistaken.

Thus the desired result was not effected, and Ireland could not be left to natural remedies; every circumstance combined in that country to exasperate the relations between landlord and tenant. The landlords were, for the most part, aliens in blood and aliens in religion. They represented conquest and confiscation, and they had gone on from generation to generation with an indifference to the welfare of the people which would not have been tolerated in England and Scotland. The law had to interfere at last to protect the peasantry in the shape of Mr Gladstone's Land Act; the best measure, perhaps the only good measure, which has been passed for Ireland for the last two hundred years.

In Ireland there are good landlords, more than are ever heard of. The object of Mr Gladstone's Act[3] was simply to shape a law out of the good landlords' practice and make the bad conform to it. It was called confiscation; I know not what was confiscated. Nothing certainly to which the landlord had any equitable right. The selling value of land has not

[3] [1870. This made eviction illegal except for non-payment of rent; it also provided compensation for improvements and loans for purchase of holdings.]

diminished in Ireland since that Act was passed. It has rather risen from the increase of security.

It is possible that a similar law may become necessary in England and Scotland. It is possible, if infinitely improbable. Responsibility is the shadow of a great position. If a time should ever come when the heirs of great estates forget that they *have* any responsibility, if they come to suppose that the world was made for them, not they for the world, that the sole duty laid on them is the duty of enjoying themselves, that they are permitted to idle away a life made weary to them by its inanity between the London season, the foreign watering-place, the deer forest, the battue, or a salmon river; then it is easy to prove that an end will come to all that.

I was staying the year before the Irish famine at a large house in Connaught. We had a great gathering there of the gentlemen of the county; more than a hundred of us sat down to a luncheon on the lawn. My neighbour at the table was a Scotchman, who was over there examining the capabilities of the soil. ' There,' he said to me, ' you see the landed gentry of this county. In all the number there may be one, at the most two, who believe that the Almighty put them into this world for any purpose but to shoot grouse, race, gamble, drink, or break their necks in the hunting-field. They are not here at all for such purposes, and one day they will find it so.'

The day of reckoning was nearer than he thought. Next year came the potato disease. The estates of most of them were mortgaged, and at best they had only a margin to live upon. Rents could not be paid. The poor people were dying of hunger, and a poor-rate had to be laid on amounting, in places, to confiscation. The Encumbered Estates Act followed, and the whole set of them were swept clean away.

We are not come to that pass here, nor do I believe we are likely to come. Even here we have heard occasionally of strange things being done; uncalled-for evictions of tenantry, with mountain and glen closed against the tourist and artist, that a noble lord and his friends may shoot a few miserable deer. But the tendency of things is not to an increase of all that, very far otherwise.

Another noble lord that I know of has a mountain property in Kerry which would make the finest deer-forest in these islands. He has the same temptation to make a deer-forest as those others have. As a forest it would bring him five times the present rent. Some forty or fifty families only would have to be removed from their farms, and they could be bought out under the Land Act with enormous profit to the landlord. But the deer are not on the mountains, and those families remain on their farms. The same noble lord spends four-fifths of the income which he draws from that property in improving the condition of the people. No one speaks of this; no one ever talks of what is done wisely and well. Health is never conscious of itself. We are only conscious of our own bodies when something is amiss with us. Offences only attract notice. Judge of British society from the police reports, and we are a nation of savages. Yet we are always forgetting this. We hear of the bad exceptions. We hear of them because they are exceptions, and we argue as if they were the rule.

Well then, gentlemen, let us turn from the mischief which may come of a landed gentry, and let us see what good comes of them.

Since land does not pay as a commercial speculation, why do rich men give such large prices for it? Land is sought after for the social consequence and for the political influence which the possession of a large estate in such a country as ours confers. It is sought after from an ambition to leave our names behind us, rooted into the soil to which the national life is attached. To obtain or keep such a position, money must be sacrificed to other considerations; and the sacrifice must be maintained and continued if the landowner is to preserve the objects for which it is made. The same force in nature appears now as heat, now as motion; one can be converted into the other. Wealth in the same way may appear in the form of luxury, or it may appear in the form of power. The landowner who desires honour and influence spends the rents which fall to him rather as a revenue than as a private income. The manager of the estates of a noble duke who is nominally one of the richest men in Great Britain said to me, that in his experience dukes never had any money.

On those estates more than a million had been laid out in a few years in rebuilding the cottages.

And the farther what is called the land monopoly is carried, the more, that is, the small estates are absorbed in the large, the better these duties will be performed. I don't know how it may be in Scotland, but I know that in England you can tell by the look of the country which you are passing through whether it belongs to a large landowner or a small one.

Compare an estate owned by one man with a hundred thousand a year, and a similar estate divided among a hundred owners with a thousand a year each. On which of these will the working tenants find themselves best off? The one great man's establishment may be expensive, but after all it is but one. The expenses of the most splendid household will not reach a hundred thousand a year, or half that sum, or a quarter of it. The great man is on a pedestal. If he is evil spoken of his pedestal becomes a pillory. Therefore he does not press his rights when he might press them. The customs of the manor are generally observed. Farm buildings are kept in good condition, fences are in good repair, cottages have roofs which will keep the rain out. You find churches, you find schools, you find everything which public opinion demands or approves.

Turn to the estate which is divided between the hundred less conspicuous proprietors. Will an equal margin of income be forthcoming for improvements? Will there be the same consideration for tenants and labourers? There cannot be, because a hundred private establishments have to be supported instead of one, and a hundred families struggling to maintain the position of gentry with inadequate means. By them every farthing which their estate will yield is required for their ordinary expenditure. They are embarrassed. They must borrow. Their obvious duties are left undone. You read the story in unmended fences, in broken gates, in decaying farm-houses. At length a crisis comes, and unless entail interferes the land is sold to some one who can better afford to keep it.

Latifundia perdidere Italiam—the great estates ruined Italy.

The yeomen who had formed the Roman armies had disappeared. The land had become the monopoly of the rich. What ruined Italy we are told will ruin Great Britain.

The argument mistakes the character of what is going on. The great estates in Italy under the empire were cultivated by slaves. The free men had been destroyed. Are the estates in Great Britain cultivated by slaves? Is the Scotch tenant who is farming another man's land a slave? Is he on the road to becoming a slave? He would be much amused if he was told so. At the bottom of his mind he knows that he is moving in an entirely opposite direction. We are but treading over again the same road which our ancestors travelled four or five centuries ago. The villein, or cultivator, under the feudal system, had originally no rights but what his lord allowed him. The lower kind of villein or serf was his lord's property as much as his horse or his dog. But custom gave the villein, by degrees, the rights of a free man. He was allowed to plead against his lord the usage of the manor. Usage passed into law, and villein tenure became copyright tenure. The English farmer became independent in all but the name, and hence grew the yeomen freeholders whose loss we are now deploring. They are gone most of them; gone because they chose to go. Look for the British copyholders now; you find them founding empires in the four quarters of the globe; but another race of them is springing from the same stem. The absolute rights of the modern landowner are slipping from his hands, with his own consent, by precisely the same process. The subtle meshes of opinion are spread over him, and landlord right submits to be restrained by reasonable tenant right.

But what the landlord loses in direct authority he gains, if he is wise, in influence, and this leads me to say a few words about countries in which a landed gentry no longer exists.

France shook off her landed proprietors at the Revolution. Many lost their heads, many more were exiled. The French landed aristocracy had become intolerable. They began to disappear of themselves. The Revolution completed what nature had commenced. France is now divided into between five and six million freeholds. At the death of a proprietor his land is shared among his children, and the partition is only arrested at the point at which the family of the cultivator can be fed. A friend of mine who wanted two or three acres for a garden had to purchase from seventeen different owners. A tenant farmer (for there are tenant farmers even in France) rents often from as many landlords as a landlord in England has tenants.

The result, undoubtedly, is thrift, industry, good spade cultivation, and great material prosperity. The magic of property, as Mr Mill long since pointed out, will turn an arid waste into a garden. The peasant works and saves because he knows that he works for himself and his family. He is conservative, for he has something of his own to lose. Were the British nation like the French, had we no colonies, and no outlet for industry at home, then a peasant proprietary might grow also in Kent and Hampshire. But what a price has France to pay for it! There is no emigration; yet the population diminishes. The law of subdivision forbids the peasant the luxury of many children. How the numbers are kept down it is needless to speculate.

While, again, a nation composed of a multitude of disconnected units is to an organised society what a heap of sand is to a block of granite, incapable of cohering for sustained political action. We shall see what the Republic can do—we are bound to wish it well—but for nearly a century France has alternated between anarchy and despotism. She tears her bonds to pieces. She allows them to be refastened when she aspires to be politically strong, and then she snaps them again when the strain becomes too violent to be borne. Never in the history of the world has any great nation been so rapidly and completely overwhelmed as France was in 1870. When her armies were defeated she had no organisation left.

The French are as public-spirited as other people, but, except under the influence of political or religious passion of a definite kind, public spirit cannot combine masses of men together for a common purpose. They have not knowledge enough, they have not confidence enough for spontaneous action. They require leaders whom they can trust, and leaders cannot be extemporised in an emergency. The natural leaders in a healthy country are the gentry; public-spirited and patriotic because their own fortunes are bound up with the fortunes of their country; personal centres of organisation because their neighbours know them, and are accustomed to look up to them. France is better without the aristocracy which she destroyed, because they were worthless. She has yet to show that she can thrive as a nation without any gentry at all.

Look again at Spain. In Spain there has been no such convulsion as the French Revolution; but in Spain, too, there is no longer any order of hereditary gentlemen. The people have not degenerated. The peasantry of Castile are as strong, as brave, as loyal as the men who followed Cortez to Mexico. Their humour is fresh as ever. Sancho Panza and his ass you may meet any day in a morning's walk, but you will find no Miguel de Cervantes and no Duke of Lerma. The tombs which lie in silent beauty in the cathedrals are all that remains of the stately Mendozas, the Olivarez, the great houses of Cordova and Toledo; and Spain is what we see. The magnificent men who three centuries ago made the Castilian monarchy the most powerful in the world have given place to eloquent orators and military adventurers. And Spain has fallen from her pride of place; her arts, her literature, her arms, once alike her glory, are now alike degraded, and the national life has perished along with them.

I have often asked myself why the hidalgos, hijos d'algo, sons of somebody, as the Cid and his comrades haughtily called themselves, have so totally disappeared. I believe it was because they did not reside on their estates among their people; because they lived in the great cities attached to the court. In Burgos, in Valladolid, in Medina, you see the places of the old nobles, their coats of arms carved in granite over the massive portals. But they had no personal relations with their tenants, or their tenants with them. They had no root, and they have withered; and have left their once-proud and glorious country the prey of priests and political charlatans and soldiers of fortune.

I shall be told that I am confounding past and present. The hidalgos are gone because they are unsuited to modern times. Public opinion, a free press, and a free platform dispense with these hereditary influences. Let the peasant and the artisan read their daily papers, and they will have no need of a gentry to lead them. It is true that much changes in this world, but there is much also which does not change, and human nature is the least changeable of all things. The English Barons extorted Magna Charta. The Long Parliament was a Parliament of English landed gentry. The English gentry made the Revolution of 1688. There is work still to be done by the descendants of those men in the country and in Parliament. Let us have all the talents in Parliament. Let trade, let science, let the learned professions, let wealth, if you like, be represented there, but it will be an ill day when we have no longer in public life the men who represent the historic traditions of Great Britain, who are returned to Parliament with no object of their own to gain, and whose services are already pledged to the commonwealth by birth and fortune.

A distinguished American once said to me, ' Hold fast to such institutions as you have left. We have none, and must do as we can without them. But do not flatter yourselves that by destroying yours you can make England like America. We are young and growing. You are in your maturity or past it. We shall rise through our difficulties. If a time comes when the English Parliament is filled with men who go there to push their own fortunes, you will perhaps not rise through yours.'

Once more. We speak contemptuously of sentiment, and yet the noblest part of our existence is based on sentiment. Patriotism is sentiment. Conscience is sentiment. Honour, shame, reverence, love of beauty, love of goodness, every high aspiration which we entertain, all are sentiment. All are unpractical according to the profit and loss philosophy. Yet without them man is but an animal, lower not higher than his fellow-creatures, as his desires are more insatiable. When I say that this question is a question of sentiment, I mean that it touches the quick of our national being.

A nation, it is said, which does not respect its past will have no future which will deserve respect. Great Britain is what it is to-day because thirty generations of strong brave men have worked with brain and hand to make it so. Nothing great ever came to men in their sleep. The fields now so clean and neatly fenced were once morasses or forests of scrub, or were littered with boulder stones. Our laws, our literature, our constitution, our empire, were built together out of materials equally unpromising. We, when we were born, came into possession of a fair inheritance. We are bound to remember from whom it came, and not to think that because we have got it we have only ourselves to thank for it. You may test the real worth of any people by the feelings which they entertain for their forefathers. With the Romans reverence for ancestors was part of the national religion. It was something like a religion here not long ago, and when the nineteenth century has sufficiently admired itself for its steam-engines and electric telegraphs, something of the same feeling, we will hope, may revive.

Every step of what is called progress for the last thousand years has been the work of some man or group of men. We talk of the tendencies of an age. The tendency of an age, unless it be a tendency to mere death and rottenness, means the energy of superior men who guide or make it; and of these superior men who have played their parts among us at successive periods the hereditary families are the monuments. Trace them back to the founders, you generally find some one whose memory ought not to be allowed to die. And usually also in the successive generations of such a family you find more than an average of high qualities, as if there was some transmission of good blood, or as if the fear of discrediting an honourable lineage was a check on folly and a stimulus to exertion. In Scotland the family histories are inseparable from the national history. How many Campbells, for instance, have not established a right to be remembered with honour? How many hundred Scotch families are there not who have produced, I will not say one distinguished man, but a whole series of distinguished men, distinguished in all branches, as soldiers, seamen, statesmen, lawyers, or men of letters?

It is true the highest names of all will not be found in the Peerages and Baronetages. The highest of all, as Burns says, take their patent of nobility direct from Almighty God.

Those patents are not made out for posterity, and the coronets which men bestow on the supremely gifted among them are usually coronets of thorns. No titled family remains as a monument of Knox or Shakespeare. They shine alone like stars. They need no monument, being themselves immortal. A Dukedom of Stratford for the descendants of Shakespeare would be like a cap and bells upon his bust. Of Knox you have not so much as a tomb—you do not know where his bones are lying. The burial-place of Knox is the heart of Protestant Scotland.

But, speaking generally, the landed gentry are enduring witnesses of past worth and good work done, and until they forfeit our esteem by demerits of their own, they deserve to be respected and honoured. High place is lost so easily that when a family has been of long continuance we may be sure that it has survived with exceptional merit. Nature rapidly finds out when the wrong sort have stolen into promotion. When a knave makes a fortune his son spends it—one generation sees an end of him. Even among the best there is a quick succession. The marble monument in the church outlasts the living one. There are no Plantagenets now; no Tudors and few Stuarts of the old stock. The Lacies and the De Courcies drop out. The Nelsons and the Wellesleys step into their places. Warriors, lawyers, politicians, press perpetually to the front. Each age has its own heroes, who in its own eyes are greater than all that went before. The worn-out material is for ever being replaced by new. Each family thus raised is on its trial. Those who survive remain as links between the present and the past, and carry on unbroken the continuity of our national existence. In such families the old expression *Noblesse oblige* is a genuine force. In a chapel attached to the church of Cheneys in Hertfordshire lies the honoured dust of ten generations of the house of Russell. There is Lord William, carried thither from the scaffold at Lincoln's Inn. There is Lady Rachel. There are the successive Earls and Dukes of Bedford, who, wise or unwise, have been always true to the people's side through three centuries of political struggle. At one end of the chapel are the monuments of the first Lord Russell, King Henry's minister at the

Reformation, and of the first Lady Russell, from whom all the rest are descended. There she lies, a stern, austere lady, as you can see in the lines of her marble countenance, evidently an exact likeness, modelled from her features. I could not but feel, as I stood in that chapel, what a thing it would be to know that in death one had to be carried into the presence of that terrible ancestress and that august array of her descendants, and to be examined whether one had been worthy of the race to which one belonged.

But enough of this, and I will bring what I have to say to an end. It appears to me, for the reasons I have given, that a landed gentry of some sort must exist in a country so conditioned as ours. The only question is whether we shall be satisfied with those that we have, or whether we wish to see them displaced in favour of others, to whom the land would, or might, be a mere commercial speculation. Abolish primogeniture, compel, either by law or by the weight of opinion, a subdivision of landed property, it will still be bought up and held in large quantities, but it will be held by successful men of business, who, being no longer able to look forward to permanence of occupancy, and therefore having no motive for wishing to secure the goodwill of the people living around them, will regard their possessions from a money point of view, and will aim at nothing but obtaining from them the largest possible amount of profit and pleasure for themselves.

A change of this kind will not conduce to our national welfare. It is perhaps coming; but I think it is still far off. The revolutionary wave which began to rise in the middle of the last century seems for the present to have spent its force. Men no longer believe that revolution will bring the millennium. They have discovered that revolution means merely a change from an aristocracy to a plutocracy, and they doubt more than they did whether much advantage comes of it after all.

The aristocracy are learning, on their side, that if they are to keep their hold in this country they must deserve to keep it. And just so far as a conviction makes its way among them that they exist for some other purpose than idle luxury, they will take out a new lease of recovered influence.

No one grudges the hard worked Member of Parliament his holidays on the moor or in the hunting-field. The days by a salmon river with the flood running off, the south-wester streaming over the pool, and the fish fresh run from the sea, are marked with chalk in the lives of the bitterest Radicals of us all. Amusement is the wine of existence, warming and feeding heart and brain. But amusement, like wine also, if taken in excess, becomes as stupid as any other form of vulgar debauchery. When we read of some noble lord, with two of his friends, shooting two thousand pheasants in a week, or that another has shot four hundred brace of partridges to his own gun in a day, we perceive that these illustrious personages have been useful to the London poulterers; but it is scarcely the work for which they are intended by the theory of their existence. The annual tournament of doves between the Lords and Commons at Hurlingham leads to odd conclusions about us on the Continent. Every institution —even the institution of a landed aristocracy—is amenable to general opinion; and it may have worse enemies than an Irish Land Act.

Fashionable follies are like soap-bubbles; the larger they are the nearer they are to bursting. Pheasant battues and pigeon-shooting will come to an end, as bull-baiting and cock-fighting came to an end. Meanwhile, the world is wide, and the British have secured handsome slices of it beyond our own island. Who in his senses—even if it were possible —would be the peasant proprietor of half a dozen acres in England when, for the sum for which he could sell them, he could buy a thousand in countries where he would be still under his own flag, among his own kindred; with an unexhausted soil, and a climate anything that he prefers, from the Arctic circle to the tropics?

You who are impatient with what you call a dependent position at home, go to Australia, go to Canada, go to New Zealand, or South Africa. There work for yourselves. There gather wealth as all but fools or sluggards are able to gather it. Come back if you will as rich men at the end of twenty years. Then buy an estate for yourselves; and when you belong to the landed gentry in your own person, you will find your eyes opened as to their value to the community.

Will you have an example of what may be done by an ordinary man with no special talents or opportunity? A Yorkshireman, an agricultural labourer, that I knew, went to Natal twelve years ago. I suppose at first he had to work for wages; and I will tell you what the wages are in that country. I stayed myself with a settler on the borders there. He had two labourers with him, an Irishman and an Englishman. They lived in his house; they fed at his own table. To the Irishman, who knew something of farming, he was paying fourteen pounds a month; to the Englishman he was paying ten; and every penny of this they were able to save.

With such wages as these, a year or two of work will bring money enough to buy a handsome property. My Yorkshireman purchased two hundred and fifty acres of wild land outside Maritzburg. He enclosed it; he carried water over it. He planted his fences with the fast-growing eucalyptus, the Australian gum-tree. In that soil and in that climate, everything will flourish, from pineapples to strawberries, from the coffee-plant and the olive to wheat and Indian corn, from oranges and bananas to figs, apples, peaches, and apricots. Now at the end of ten years the mere gum-trees which I saw on that man's land could be sold for two thousand pounds, and he is making a rapid fortune by supplying fruit and vegetables to the market at Maritzburg.

Here, as it seems to me, is the true solution of the British land question. What a Yorkshireman can do I suppose a Scotchman can do. There is already a new Scotland, so called, in South Africa; a land of mountains and valleys and rocky streams and rolling pastures. And there is gold there, and coal, and iron, and all the elements of wealth. People that country, people any part of any of our colonies, from the younger sons who complain that there is no room for them at home. Match the New England across the Atlantic with a New Scotland in South Africa; only tie it tighter to the old country. Spread out there and everywhere. Take possession of the boundless inheritance which is waiting for you, and leave the old Island to preserve its ancient memories under such conditions as the times permit.

Amidst the varied reflections which the nineteenth century
is in the habit of making on its condition and its prospects,
there is one common opinion in which all parties coincide—
that we live in an era of progress. Earlier ages, however
energetic in action, were retrospective in their sentiments.
The contrast between a degenerate present and a glorious past
was the theme alike of poets, moralists, and statesmen. When
the troubled Israelite demanded of the angel why the old
times were better than the new, the angel admitted the fact
while rebuking the curiosity of the questioner. ' Ask not the
cause,' he answered. ' Thou does not inquire wisely concerning
this.' As the hero of Nestor's youth flung the stone with ease
which twelve of the pigmy chiefs before Troy could scarcely
lift from the ground, so ' the wisdom of our ancestors ' was
the received formula for ages with the English politician.
Problems were fairly deemed insoluble which had baffled
his fathers, ' who had more wit and wisdom than he.' We
now know better, or we imagine that we know better, what the
past really was. We draw comparisons, but rather to encourage
hope than to indulge despondency or foster a deluding
reverence for exploded errors. The order of the ages is
inverted. Stone and iron came first. We ourselves may
possibly be in the silver stage. An age of gold, if the
terms of our existence on this planet permit the contemplation
of it as a possibility, lies unrealized in the future. Our lights
are before us, and all behind is shadow. In every department
of life—in its business and in its pleasures, in its beliefs and
in its theories, in its material developments and in its spiritual
convictions—we thank God that we are not like our fathers.
And while we admit their merits, making allowance for
their disadvantages, we do not blind ourselves in mistaken
modesty to our own immeasurable superiority.

Changes analogous to those which we contemplate with
so much satisfaction have been witnessed already in the history
of other nations. The Roman in the time of the Antonines

might have looked back with the same feelings on the last years of the Republic. The civil wars were at an end. From the Danube to the African deserts, from the Euphrates to the Irish Sea, the swords were beaten into ploughshares. The husbandman and the artisan, the manufacturer and the merchant, pursued their trades under the shelter of the eagles, secure from arbitrary violence, and scarcely conscious of their masters' rule. Order and law reigned throughout the civilized world. Science was making rapid strides. The philosophers of Alexandria had tabulated the movements of the stars, had ascertained the periods of the planets, and were anticipating by conjecture the great discoveries of Copernicus. The mud cities of the old world were changed to marble. Greek art, Greek literature, Greek enlightenment, followed in the track of the legions. The harsher forms of slavery were modified. The bloody sacrifices of the Pagan creeds were suppressed by the law; the coarser and more sensuous superstitions were superseded by a broader philosophy. The period between the accession of Trajan and the death of Marcus Aurelius has been selected by Gibbon as the time in which the human race had enjoyed more general happiness than they had ever known before, or had known since, up to the date when the historian was meditating on their fortunes. Yet during that very epoch, and in the midst of all that prosperity, the heart of the empire was dying out of it. The austere virtues of the ancient Romans were perishing with their faults. The principles, the habits, the convictions, which held society together were giving way, one after the other, before luxury and selfishness. The entire organization of the ancient world was on the point of collapsing into a heap of incoherent sand.

If the merit of human institutions is at all measured by their strength and stability, the increase of wealth, of production, of liberal sentiment, or even of knowledge, is not of itself a proof that we are advancing on the right road. The unanimity of the belief therefore that we are advancing at present must be taken as a proof that we discern something else than this in the changes which we are undergoing. It would be well, however, if we could define more clearly what we precisely do discern. It would at once be a relief to the weaker brethren whose minds occasionally misgive them, and it would throw out into distinctness the convictions which we have at length arrived at on the true constituents of human worth, and the objects towards which human beings ought to direct their energies. We are satisfied that we are going forward. That is to be accepted as no longer needing proof. Let us ascertain or define in what particulars and in what direction we are going forward, and we shall then understand in what improvement really consists.

The question ought not to be a difficult one, for we have abundant and varied materials. The advance is not confined to ourselves. France, we have been told any time these twenty years, has been progressing enormously under the beneficent rule of Napoleon III. Lord Palmerston told us, as a justification of the Crimean war, that Turkey had made more progress in the two preceding generations than any country in the world. From these instances we might infer that Progress was something mystic and invisible, like the operation of the graces said to be conferred in baptism. The distinct idea which was present in Lord Palmerston's mind is difficult to discover. In the hope that some enlightened person will clear up an obscurity which exists only perhaps in our own want of perception, I proceed to mention some other instances in which, while I recognize change, I am unable to catch the point of view from which to regard it with unmixed satisfaction. Rousseau maintained that the primitive state of man was the happiest, that civilization was corruption, and that human nature deteriorated with the complication of the conditions of its existence. A paradox of that kind may be defended as an entertaining speculation. I am not concerned with any such barren generalities. Accepting social organization as the school of all that is best in us, I look merely to the alterations which it is undergoing; and if in some things passing away it seems to me that we are lightly losing what we shall miss when they are gone and cannot easily replace, I shall learn gladly that I am only suffering under the proverbial infirmity of increasing years, and that, like Esdras, I perplex myself to no purpose.

Let me lightly, then, run over a list of subjects on which the believer in progress will meet me to most advantage.

I

I will begin with the condition of the agricultural poor, the relation of the labourer to the soil, and his means of subsistence.

The country squire of the last century, whether he was a Squire Western or a Squire Allworthy, resided for the greater part of his life in the parish where he was born. The number of freeholders was four times what it is at present; plurality of estates was the exception; the owner of land, like the peasant, was virtually *ascriptus glebæ*—a practical reality in the middle of the property committed to him. His habits, if he was vicious, were coarse and brutal—if he was a rational being, were liberal and temperate; but in either case the luxuries of modern generations were things unknown to him. His furniture was massive and enduring. His household expenditure, abundant in quantity, provided nothing of the costly delicacies which it is now said that every one expects and every one therefore feels bound to provide. His son at Christ-church[1] was contented with half the allowance which a youth with expectations now holds to be the least on which he can live like a gentleman. His servants were brought up in the family as apprentices, and spent their lives under the same roof. His wife and his daughters made their own dresses, darned their own stockings, and hemmed their own handkerchiefs. The milliner was an unknown entity at houses where the milliner's bill has become the unvarying and not the most agreeable element of Christmas. A silk gown lasted a lifetime, and the change in fashions was counted rather by generations than by seasons. A London house was unthought of—a family trip to the Continent as unimaginable as an outing to the moon. If the annual migration was something farther than, as in Mr Primrose's parsonage,[2] from the blue room to the brown,

[1] [The college at Oxford.]
[2] [In Goldsmith's *Vicar of Wakefield.*]

it was limited to the few weeks at the county town. Enjoyments were less varied and less expensive. Home was a word with a real meaning. Home occupations, home pleasures, home associations and relationships, filled up the round of existence. Nothing else was looked for, because nothing else was attainable. Among other consequences, habits were far less expensive. The squire's income was small as measured by modern ideas. If he was self-indulgent, it was in pleasures which lay at his own door, and his wealth was distributed among those who were born dependent on him. Every family on the estate was known in its particulars, and had claims for consideration which the better sort of gentlemen were willing to recognize. If the poor were neglected, their means of taking care of themselves were immeasurably greater than at present. The average squire may have been morally no better than his great-grandson. In many respects he was probably worse. He was ignorant, he drank hard, his language was not particularly refined, but his private character was comparatively unimportant; he was controlled in his dealings with his people by the traditional English habits which had held society together for centuries—habits which, though long gradually decaying, have melted entirely away only within living memories.

At the end of the sixteenth century an Act passed obliging the landlord to attach four acres of land to every cottage on his estate. The act itself was an indication that the tide was on the turn. The English villein, like the serf all over Europe, had originally rights in the soil, which were only gradually stolen from him. The statute of Elizabeth was a compromise reserving so much of the old privileges as appeared indispensable for a healthy life.

The four acres shrivelled like what had gone before; but generations had to pass before they had dwindled to nothing, and the labourer was inclosed between his four walls to live upon his daily wages.

Similarly, in most country parishes there were tracts of common land, where every householder could have his flock of sheep, his cow or two, his geese or his pig; and milk and bacon so produced went into the limbs of his children, and went to form the large English bone and sinew which are becoming things of tradition. The thicket or the peat bog provided fuel. There were spots where the soil was favourable in which it was broken up for tillage, and the poor families in rotation raised a scanty crop there. It is true that the common land was wretchedly cultivated. What is every one's property is no one's property. The swamps were left undrained, the gorse was not stubbed up. The ground that was used for husbandry was racked. An inclosed common taken in hand by a man of capital produces four, five, or six times what it produced before. But the landlord who enters on possession is the only gainer by the change. The cottagers made little out of it, but they made something, and that something to them was the difference between comfort and penury. The inclosed land required some small additional labour. A family or two was added to the population on the estate, but it was a family living at the lower level to which all had been reduced. The landlord's rent-roll shows a higher figure, or it may be he has only an additional pheasant preserve. The labouring poor have lost the faggot on their hearths, the milk for their children, the slice of meat at their own dinners.

Even the appropriation of the commons has not been sufficient without closer paring. When the commons went, there was still the liberal margin of grass on either side of the parish roads, to give pickings to the hobbled sheep or donkey. The landlord, with the right of the strong, which no custom can resist, is now moving forward his fences, taking possession of these ribands of green, and growing solid crops upon them. The land is turned to better purpose. The national wealth in some inappreciable way is supposed to have increased, but the only visible benefit is to the lord of the soil, and appears in some added splendour to the furniture of his drawing-room.

It is said that men are much richer than they were, that luxury is its natural consequence, and is directly beneficial to the community as creating fresh occupations and employing more labour. The relative produce of human industry, however, has not materially increased in proportion to the growth of population. ' If riches increase, they are increased that eat them.' If all the wealth which is now created in this country was distributed among the workers in the old ratio, the margin which could be spent upon personal self-indulgence would not be very much larger than it used to be. The economists insist that the growth of artificial wants among the few is one of the symptoms of civilization—is a means provided by nature to spread abroad the superfluities of the great. If the same labour, however, which is now expended in the decorating and furnishing a Belgravian palace was laid out upon the cottages on the estates of its owner, an equal number of workmen would find employment, an equal fraction of the landlords' income would be divided in wages. For the economist's own purpose, the luxury could be dispensed with if the landlord took a different view of the nature of his obligations. Progress and civilization conceal the existence of his obligations, and destroy at the same time the old-fashioned customs which limited the sphere of his free will. The great estates have swallowed the small. The fat ears of corn have eaten up the lean. The same owner holds properties in a dozen counties. He cannot reside upon them all, or make personal acquaintance with his multiplied dependants. He has several country residences. He lives in London half the year, and most of the rest upon the Continent. Inevitably he comes to regard his land as an investment; his duty to it the development of its producing powers; the receipt of his rents the essence of the connection; and his personal interest in it the sport which it will provide for himself and his friends. Modern landlords frankly tell us that if the game laws are abolished, they will have lost the last temptation to visit their country seats. If this is their view of the matter, the sooner they sell their estates and pass them over to others, to whom life has not yet ceased to be serious, the better it will be for the community. They complain of the growth of democracy and insubordination. The fault is wholly in themselves. They have lost the respect of the people because they have ceased to deserve it.

11

If it be deemed a paradox to maintain that the relation between the owners of land and the peasantry was more satisfactory in the old days than in the present, additional hardiness is required to assert that there has been no marked improvement in the clergy. The bishop, rector, or vicar of the Established Church in the eighteenth century is a by-word in English ecclesiastical history. The exceptional distinction of a Warburton or a Wilson, a Butler or a Berkeley, points the contrast only more vividly with the worldliness of their brothers on the bench. The road to honours was through political sub-serviency. The prelates indemnified themselves for their ignominy by the abuse of their patronage, and nepotism and simony were too common to be a reproach. Such at least is the modern conception of these high dignitaries, which instances can be found to justify. In an age less inflated with self-esteem, the nobler specimens would have been taken for the rule, the meaner and baser for the exception. Enough, however, can be ascertained to justify the enemies of the Church in drawing an ugly picture of the condition of the hierarchy. Of the parochial clergy of those times the popular notion is probably derived from Fielding's novels. Parson Trulliber is a ruffian who would scarcely find admittance into a third-rate farmers' club of the present day. Parson Adams, a low-life Don Quixote, retains our esteem for his character at the expense of contempt for his understanding. The best of them appear as hangers-on of the great, admitted to a precarious equality in the housekeeper's room, their social position being something lower than that of the nursery governess in the establishment of a vulgar millionaire.

That such specimens as these were to be found in England in the last century is no less certain than that in some parts of the country the type may be found still surviving. That they were as much exceptions we take to be equally clear. Those who go for information to novels may remember that

there was a Yorick as well as a Phutatorius or a Gastripheres.[8]
Then, more than now, the cadets of the great houses were
promoted, as a matter of course, to the family livings, and
were at least gentlemen. Sydney Smith's great prizes of the
Church were as much an object of ambition to men of birth as
the high places in the other professions; and between
pluralities and sinecures, cathedral prebendaries, and the
fortunate possessors of two or more of the larger benefices,
held their own in society with the county families, and lived
on equal terms with them. If in some places there was
spiritual deadness and slovenliness, in others there was
energy and seriousness. Clarissa Harlowe found daily service
in the London churches as easily as she could find it now.

That the average character of the country clergy, however,
was signally different from what it is at present, is not to be
disputed. They were Protestants to the backbone. They knew
nothing and cared nothing about the Apostolical Succession.
They had no sacerdotal pretensions; they made no claims to be
essentially distinguished from the laity. Their official duties
sat lightly on them. They read the Sunday services,
administered the Communion four times a year, preached
commonplace sermons, baptized the children, married them
when they grew to maturity, and buried them when they died;
and for the rest they lived much as other people lived, like
country gentlemen of moderate fortune, and, on the whole,
setting an example of respectability. The incumbents of
benefices over a great part of England were men with small
landed properties of their own. They farmed their own
glebes. They were magistrates, and attended quarter sessions
and petty sessions, and in remote districts, where there were no
resident gentry of consequence, were the most effective
guardians of the public peace. They affected neither austerity
nor singularity. They rode, shot, hunted, ate and drank, like
other people; occasionally, when there was no one else to
take the work upon them, they kept the hounds. In dress and
habit they were simply a superior class of small country
gentlemen; very far from immaculate, but, taken altogether,

[8] [These names of imaginary characters occur in the novels of
Laurence Sterne.]

wholesome and solid members of practical English life. It may seem like a purposed affront to their anxious and pallid successors, clad in sacerdotal uniform, absorbed in their spiritual functions, glorying in their Divine commission, passionate theologians, occupied from week's end to week's end with the souls of their flocks, to contrast them unfavourably with secular parsons who, beyond their mechanical offices, had nothing of the priest to distinguish them; yet it is no less certain that the rector of the old school stood on sounder terms with his parishioners, and had stronger influence over their conduct. He had more in common with them. He understood them better, and they understood him better. The Establishment was far more deeply rooted in the affections of the people. The measure of its strength may be found in those very abuses, so much complained of, which, nevertheless, it was able to survive. The forgotten toast of Church and King was a matter of course at every county dinner. The omission of it would have been as much a scandal as the omission of grace. Dissenters sat quiescent under disabilities which the general sentiment approved. The revival of spiritual zeal has been accompanied with a revival of instability. As the clergy have learnt to magnify their office, the laity have become indifferent or hostile.

Many causes may be suggested to explain so singular a phenomenon. It is enough to mention one. The parson of the old school, however ignorant of theology, however outwardly worldly in character, did sincerely and faithfully believe in the truth of the Christian religion; and the congregation which he addressed was troubled with as few doubts as himself. Butler and Berkeley speak alike of the spread of infidelity; but it was an infidelity confined to the cultivated classes—to the London wits who read Bolingbroke or Hume's *Essays* or *Candide*. To the masses of the English people, to the parishioners who gathered on Sundays into the churches, whose ideas were confined to the round of their common occupations, who never left their own neighbourhood, never saw a newspaper or read a book but the Bible and the *Pilgrim's Progress*, the main facts of the Gospel history were as indisputably true as the elementary laws of the universe.

That Christ had risen from the dead was as sure as that the sun had risen that morning. That they would themselves rise was as certain as that they would die; and as positively would one day be called to judgment for the good or ill that they had done in life. It is vain to appeal to their habits as a proof that their faith was unreal. Every one of us who will look candidly into his own conscience can answer that objection. Every one of us, whatever our speculative opinions, knows better than he practises, and recognizes a better law than he obeys. Belief and practice tend in the long run, and in some degree, to correspond; but in detail and in particular instances they may be wide asunder as the poles. The most lawless boys at school, and the loosest young men at college, have the keenest horror of intellectual scepticism. Their passions may carry them away; but they look forward to repenting in the end. Later in life they may take refuge in infidelity if they are unable to part with their vices; but the compatibility of looseness of habit with an unshaken conviction of the general truths of religion is a feature of our nature which history and personal experience alike confirm.

It is unnecessary to dwell upon the change which has passed over us all during the last forty years. The most ardent ritualist now knows at heart that the ground is hollow under him. He wrestles with his uncertainties. He conceals his misgivings from his own eyes by the passion with which he flings himself into his work. He recoils, as every generous-minded man must recoil, from the blankness of the prospect which threatens to open before him. To escape the cloud which is gathering over the foundations of his faith he busies himself with artificial enthusiasm in the external expressions of it. He buries his head in his vestments. He is vehement upon doctrinal minutiæ, as if only these were at stake. He clutches at the curtains of mediæval theology to hide his eyes from the lightning which is blinding him. His efforts are vain. His own convictions are undermined in spite of him. What men as able as he is to form an opinion doubt about, by the nature of the case is made doubtful. And neither in himself nor in the congregations whom he adjures so passionately is there any basis of unshaken belief remaining.

He is like a man toiling with all his might to build a palace out of dry sand. Ecclesiastical revivals are going on all over the world, and all from the same cause. The Jew, the Turk, the Hindoo, the Roman Catholic, the Anglo-Catholic, the Protestant English Dissenter, are striving with all their might to blow into flame the expiring ashes of their hearth fires. They are building synagogues and mosques, building and restoring churches, writing books and tracts; persuading themselves and others with spasmodic agony that the thing they love is not dead, but sleeping. Only the Germans, only those who have played no tricks with their souls, and have carried out boldly the spirit as well as the letter of the Reformation, are meeting the future with courage and manliness, and retain their faith in the living reality while the outward forms are passing away.

III

The Education question is part of the Church question, and we find in looking at it precisely the same phenomena. Education has two aspects. On one side it is the cultivation of man's reason, the development of his spiritual nature. It elevates him above the pressure of material interests. It makes him superior to the pleasures and the pains of a world which is but his temporary home, in filling his mind with higher subjects than the occupations of life would themselves provide him with. One man in a million of peculiar gifts may be allowed to go no farther, and may spend his time in pursuits merely intellectual. A life of speculation to the multitude, however, would be a life of idleness and uselessness. They have to maintain themselves in industrious independence in a world in which it has been said there are but three possible modes of existence, begging, stealing, and working; and education means also the equipping a man with means to earn his own living. Every nation which has come to anything considerable has grown by virtue of a vigorous and wholesome education. A nation is but the aggregate of the individuals of which it is composed. Where individuals grow up ignorant and incapable, the result is anarchy and torpor. Where there has been energy, and organized strength, there is or has been also an effective training of some kind. From a modern platform speech one would infer that before the present generation the schoolmaster had never been thought of, and that the English of past ages had been left to wander in darkness. Were this true, they would have never risen out of chaos. The problem was understood in Old England better probably than the platform orator understands it, and received a more practical solution than any which on our new principles has yet been arrived at. Five out of six of us have to earn our bread by manual labour, and will have to earn it so to the end of the chapter. Five out of six English children in past generations were in consequence apprenticed' to some trade or calling by which that necessary feat could be surely accomplished. They learnt in their catechisms and at church that they were responsible to their Maker for the use which they made of their time. They were taught that there was an immortal part of them, the future of which depended on their conduct while they remained on earth. The first condition of a worthy life was to be able to live honestly; and in the farm or at the forge, at the cobbler's bench or in the carpenter's yard, they learnt to stand on their own feet, to do good and valuable work for which society would thank and pay them. Thenceforward they could support themselves and those belonging to them without meanness, without cringing, without demoralizing obligation to others, and had laid in rugged self-dependence the only foundation for a firm and upright character. The old English education was the apprentice system. In every parish in England the larger householders, the squire and the parson, the farmers, smiths, joiners, shoemakers, were obliged by law to divide among themselves according to their means the children of the poor who would otherwise grow up unprovided for, and clothe, feed, lodge, and teach them in return for their services till they were old enough to take care of themselves. This was the rule which was acted upon for many centuries. It broke down at last. The burden was found disagreeable; the inroad too heavy upon natural ability. The gentlemen were the first to decline or evade their obligations. Their business was to take boys and girls for household service. They preferred to have their servants ready made. They did not care to encumber their establishments with awkward urchins or untidy slatterns, who broke their china and whom they were unable to dismiss. The farmers and the artisans objected naturally to bearing the entire charge—they who had sufficient trouble to keep their own heads above water : they had learnt from the gentlemen that their first duties were to themselves, and their ill humour vented itself on the poor little wretches who were flung upon their unwilling hands. The children were ill-used, starved, beaten. In some instances they were killed. The benevolent instincts of the country took up their cause. The apprenticeship under its compulsory form passed away amidst universal execrations. The masters were relieved from the obligation to educate, the lads themselves from the obligation to be educated. They were left to their parents, to their own helplessness, to the chances and casualties of life, to grow up as they could, and drift untaught into whatever occupation they could find. Then first arose the cry of the schoolmaster. The English clergy deserve credit for having been the first to see the michief that must follow, and to look for a remedy. If these forlorn waifs and strays could no longer be trained, they could not be permitted to become savages. They could learn, at least, to read and write. They could learn to keep themselves clean. They could be broken into habits of decency and obedience, and be taught something of the world into which they were to be flung out to sink or swim. Democracy gave an impulse to the movement. 'We must educate our masters,' said Mr Lowe[4] sarcastically. Whether what is now meant by education will make their rule more intelligent remains to be seen. Still the thing is to be done. Children whose parents cannot help them are no longer utterly without a friend. The State charges itself with their

[4] [Robert Lowe, first Viscount Sherbrooke, (1811-1892), served under Gladstone as Chancellor of the Exchequer and Home Secretary, 1868-1874.]

minds, if not their bodies. Henceforward they are to receive such equipment for the battle of life as the schoolmaster can provide.

It is something, but the event only can prove that it will be as useful as an apprenticeship to a trade, with the Lord's Prayer and the Commandments at its back. The conditions on which we have our being in this planet remain unchanged. Intelligent work is as much a necessity as ever, and the proportion of us who must set our hands to it is not reduced. Labour is the inevitable lot of the majority, and the best education is that which will make their labour most productive. I do not undervalue book knowledge. Under any aspect it is a considerable thing. If the books be well chosen and their contents really mastered, it may be a beautiful thing; but the stubborn fact will remain, that after the years, be they more or be they less, which have been spent at school, the pupil will be launched into life as unable as when he first entered the school door to earn a sixpence, possessing neither skill nor knowledge for which any employer in England will be willing to hire his services. An enthusiastic clergyman who had meditated long on the unfairness of confining mental culture to the classes who had already so many other advantages, gave his village boys the same education which he had received himself. He taught them languages and literature, and moral science, and art and music. He unfitted them for the state of life in which they were born. He was unable to raise them into a better. He sent one of the most promising of them with high recommendations to seek employment in a London banking-house. The lad was asked what he could do. It was found that, allowing for his age, he could pass a fair examination in two or three plays of Shakespeare.

Talent, it is urged, real talent, crippled hitherto by want of opportunity, will be enabled to show itself. It may be so. Real talent, however, is not the thing which we need be especially anxious about. It can take care of itself. If we look down the roll of English worthies in all the great professions, in church and law, in army and navy, in literature, science and trade, we see at once that the road must have been always open for boys of genius to rise. We have to consider the million, not the units; the average, not the exceptions.

It is argued again that by educating boys' minds, and postponing till later their special industrial training, we learn better what each is fit for; time is left for special fitnesses to show themselves. We shall make fewer mistakes, and boys will choose the line of life for which nature has qualified them. This may sound plausible, but capacity of a peculiarly special kind is the same as genius, and may be left to find its own place. A Canova or a Faraday makes his way through all impediments into the occupation which belongs to him. Special qualifications, unless they are of the highest order, do not exist to a degree worth considering. A boy's nature runs naturally into the channel which is dug for it. Teach him to do any one thing, and in doing so you create a capability; and you create a taste along with it; his further development will go as far and as wide as his strength of faculty can reach; and such varied knowledge as he may afterwards accumulate will grow as about a stem round the one paramount occupation which is the business of his life.

A sharp lad, with general requirements, yet unable to turn his hand to one thing more than another, drifts through existence like a leaf blown before the wind. Even if he retains what he has learnt, it is useless to him. The great majority so taught do not retain, and cannot retain, what they learn merely as half-understood propositions, and which they have no chance of testing by practice. Virgil and Sophocles, logic and geometry, with the ordinary university pass-man, are as much lost to him in twenty years from his degree as if he had never construed a line or worked a problem. Why should we expect better of the pupil of the middle or lower class, whose education ends with his boyhood? Why should his memory remain burdened with generalities of popular science, names and dates from history which have never been more than words to him, or the commonplaces of political economy, which, if he attaches any meaning at all to them, he regards as the millionaire's catechism, which he will believe when he is a millionaire himself? The knowledge which a man can use is the only real knowledge, the only knowledge which has life and growth in it, and converts itself into practical power. The rest hangs like dust about the brain, or dries like raindrops off the stones.

The mind expands, we are told; larger information generates larger and nobler thoughts. Is it so? We must look to the facts. General knowledge means general ignorance, and an ignorance, unfortunately, which is unconscious of itself. Quick wits are sharpened up. Young fellows so educated learn that the world is a large place, and contains many pleasant things for those who can get hold of them. Their ideas doubtless are inflated, and with them their ambitions and desires. They have gained nothing towards the wholesome gratifying of those desires, while they have gained considerable discontent at the inequalities of what is called fortune. They are the ready-made prey of plausible palaver written or spoken, but they are without means of self-help, without seriousness, and without stability. They believe easily that the world is out of joint because they, with their little bits of talents, miss the instant recognition which they think their right. Their literature, which the precious art of reading has opened out to them, is the penny newspaper; their creed, the latest popular chimera which has taken possession of the air. They form the classes which breed like mushrooms in the modern towns, and are at once the scorn and the perplexity of the thoughtful statesman. They are Fenians in Ireland, trades-unionists in England, rabid partisans of slavery or rabid abolitionists in America, socialists and red republicans on the Continent. It is better that they should have any education than none. The evils caused by a smattering of information, sounder knowledge may eventually cure. I refuse only to admit that the transition from the old industrial education to the modern book education is, for the present or the immediate future, a sign of what can be called progress.

Let there be more religion, men say. Education will not do without religion. Along with the secular lessons we must have Bible lessons, and then all will go well. It is perfectly true that a consciousness of moral responsibility, a sense of the obligation of truth and honesty and purity, lies at the bottom of all right action—that without it knowledge is useless, that with it everything will fall into its place. But it is with religion as with all else of which I am speaking. Religion can be no more learnt out of books than seamanship, or soldiership, or engineering, or painting, or any practical trade whatsoever. The doing right alone teaches the value or the meaning of right; the doing it willingly, if the will is happily constituted; the doing it unwillingly, or under compulsion, if persuasion fails to convince. The general lesson lies in the commandment once taught with authority by the clergyman; the application of it in the details of practical life, in the execution of the particular duty which each moment brings with it. The book lesson, be it Bible lesson, or commentary, or catechism, can at best be nothing more than the communication of historical incidents of which half the educated world have begun to question the truth, or the dogmatic assertion of opinions over which theologians quarrel and will quarrel to the end of time. France has been held up before us for the last twenty years as the leader of civilization, and Paris as the head-quarters of it. The one class in this supreme hour of trial for that distracted nation in which there is most hope of good is that into which the ideas of Paris have hitherto failed to penetrate. The French peasant sits as a child at the feet of the priesthood of an exploded idolatry. His ignorance of books is absolute; his superstitions are contemptible; but he has retained a practical remembrance that he has a Master in Heaven who will call him to account for his life. In the cultivation of his garden and vineyard, in the simple round of agricultural toil, he has been saved from the temptation of the prevailing delusions, and has led, for the most part, a thrifty, self-denying, industrious, and useful existence. Keener sarcasm it would be hard to find on the inflated enthusiasm of progress.

IV

Admitting—and we suspect very few of our readers will be inclined to admit—that there is any truth in these criticisms, it will still be said that our shortcomings are on the way to cure themselves. We have but recently roused ourselves from past stagnation, and that a new constitution of things cannot work at once with all-sided perfection is no more than we might expect. Shortcomings there may be, and our business is to find them out and mend them. The means are now in our hands. The people have at last political power. All interests are now represented in Parliament. All are sure of consideration. Class government is at an end. Aristocracies, landowners, established churches, can abuse their privileges no longer. The age of monopolies is gone. England belongs to herself. We are at last free.

It would be well if there were some definition of freedom which would enable men to see clearly what they mean and do not mean by that vaguest of words. The English Liturgy says that freedom is to be found perfectly in the service of God. ' *Intellectual emancipation,*' says Goethe, ' *if it does not give us at the same time control over ourselves, is poisonous.*' Undoubtedly the best imaginable state of human things would be one in which everybody thought with perfect correctness and acted perfectly well of his own free will, unconstrained, and even unguided, by external authority. But inasmuch as no such condition as this can be looked for this side of the day of judgment, the question for ever arises how far the unwise should be governed by the wise—how far society should be protected against the eccentricities of fools, and fools be protected against themselves. There is a right and a wrong principle on which each man's life can be organized. There is a right or a wrong in detail at every step he takes. Much of this he must learn for himself. He must learn to act as he learns to walk. He obtains command of his limbs by freely using them. To hold him up each time that he totters is to deprive him of his only means of learning how not to fall. There are other things in which it is equally clear that he must not be left to himself. Not only may he not in the exercise of his liberty do what is injurious to others—he must not seriously injure himself. A stumble or a fall is a wholesome lesson to take care, but he is not left to learn by the effects that poison is poison, or getting drunk is brutalizing. He is forbidden to do what wiser men than he know to be destructive to him.

If he refuses to believe them, and acts on his own judgment, he is not gaining any salutary instruction—he is simply hurting himself, and has a just ground of complaint ever after against those who ought to have restrained him. As we ' become our own masters,' to use the popular phrase, we are left more and more to our own guidance, but we are never so entirely masters of ourselves that we are free from restraint altogether. The entire fabric of human existence is woven of the double threads of freedom and authority, which are for ever wrestling one against the other. Their legitimate spheres slide insensibly one into the other. The limits of each vary with time, circumstance, and character, and no rigid line can be drawn which neither ought to overpass. There are occupations in which error is the only educator. There are actions which it is right to blame, but not forcibly to check or punish. There are actions, again—actions like suicide—which may concern no one but a man's self, yet which nevertheless it may be right forcibly to prevent. Precise rules cannot be laid down which will meet all cases.

The private and personal habits of grown men lie for the most part outside the pale of interference. It is otherwise, however, in the relations of man to society. There, running through every fibre of those relations, is justice and injustice —justice which means the health and life of society, injustice which is poison and death. As a member of society a man parts with his natural rights, and society in turn incurs a debt to him which it is bound to discharge. Where the debt is adequately rendered, where on both sides there is a consciousness of obligation, where rulers and ruled alike understand that more is required of them than attention to their separate interests, and where they discern with clearness in what that ' more ' consists, there at once is good government, there is supremacy of law—law written in the statute book, and law written in the statute book of Heaven; and there, and only there, is freedom.

Das Gesetz soll nur uns Freiheit geben.[5]

As in personal morality liberty is self-restraint, and self-indulgence is slavery, so political freedom is possible only

[5] [' Only Law will give us freedom.']

where justice is in the seat of authority, where all orders and degrees work in harmony with the organic laws which man neither made nor can alter—where the unwise are directed by the wise, and those who are trusted with power use it for the common good.

A country so governed is a free country, be the form of the constitution what it may. A country not so governed is in bondage, be its suffrage never so universal. Where justice is supreme, no subject is forbidden anything which he has a right to do or to desire; and therefore it is that political changes, revolutions, reforms, transfers of power from one order to another, from kings to aristocracies, from aristocracies to peoples, are in themselves no necessary indications of political or moral advance. They mean merely that those in authority are no longer fit to be trusted with exclusive power. They mean that those high persons are either ignorant and so incapable, or have forgotten the public good in their own pleasures, ambitions, or superstitions; that they have ceased to be the representatives of any superior wisdom or deeper moral insight, and may therefore justly be deprived of privileges which they abuse for their own advantage and for public mischief. Healthy nations when justly governed never demand constitutional changes. Men talk of entrusting power to the people as a moral education, as enlarging their self-respect, elevating their imaginations, making them alive to their dignity as human beings. It is well, perhaps, that we should dress up in fine words a phenomenon which is less agreeable in his nakedness. But at the bottom of things the better sort are always loyal to governments which are doing their business well and impartially. They doubt the probability of being themselves likely to mend matters, and are thankful to let well alone. The growth of popular constitutions in a country originally governed by an aristocracy implies that the aristocracy is not any more a real aristocracy—that it is alive to its own interests and blind to other people's interests. It does not imply that those others are essentially wiser or better, but only that they understand where their own shoe pinches; and that if it be merely a question of interest, they have a right to be considered as well as the class above them. In one sense it may be called an advance, that in the balance of power so introduced particular forms of aggravated injustice may be rendered impossible; but we are brought no nearer to the indispensable thing without which no human society can work healthily or happily—the sovereignty of wisdom over folly—the pre-eminence of justice and right over greediness and self-seeking. The unjust authority is put away, the right authority is not installed in its place. People suppose it a great thing that every English householder should have a share in choosing his governors. Is it that the functions of government being reduced to a cypher, the choice of its administrators may be left to haphazard? The crew of a man-of-war understand something of seamanship; the rank and file of a regiment are not absolutely without an inkling of the nature of military service; yet if seamen and soldiers were allowed to choose their own leaders, the fate of fleets and armies so officered would not be hard to predict. Because they are not utterly ignorant of their business, and because they do not court their own destruction, the first use which the best of them would make of such a privilege would be to refuse to act upon it.

No one seriously supposes that popular suffrage gives us a wiser Parliament than we used to have. Under the rotten borough system Parliament was notoriously a far better school of statesmanship than it is or ever can be where the merits of candidates have first to be recognized by constituencies. The rotten borough system fell, not because it was bad in itself, but because it was abused to maintain injustice—to enrich the aristocracy and the landowners at the expense of the people. We do not look for a higher morality in the classes whom we have admitted to power; we expect them only to be sharp enough to understand their own concerns. We insist that each interest shall be represented, and we anticipate from the equipoise the utmost attainable amount of justice. It may be called progress, but it is a public confession of despair of human nature. It is as much as to say, that although wisdom may be higher than folly as far as heaven is above earth, the wise man has no more principle than the fool. Give him power and he will read the moral laws of the universe into a code which will only fill his own pocket, and being no better than the fool, has no more right to be listened to. The entire Civil Service of this country has been opened amidst universal acclamations to public competition. Any one who is not superannuated, and has not incurred notorious disgrace, may present himself to the Board of Examiners, and win himself a place in a public department. Everybody knows that if the heads of the departments were honestly to look for the fittest person that they could find to fill a vacant office, they could make better selections than can be made for them under the new method. The alteration means merely that these superior persons will not or cannot use their patronage disinterestedly, and that of two bad methods of choice the choice by examination is the least mischievous.

The world calls all this progress. I call it only change; change which may bring us nearer to a better order of things, as the ploughing up and rooting the weeds out of a fallow is a step towards growing a clean crop of wheat there, but without a symptom at present showing of healthy organic growth. When a block of type from which a book has been printed is broken up into its constituent letters the letters so disintegrated are called ' pie.' The pie, a mere chaos, is afterwards sorted and distributed, preparatory to being built up into fresh combinations. A distinguished American friend describes Democracy as ' making pie.'

Meanwhile, beside the social confusion, the knowledge of outward things and the command of natural forces are progressing really with steps rapid, steady, and indeed gigantic. ' Knowledge comes ' if ' wisdom lingers.' The man of science discovers; the mechanist and the engineer appropriate and utilize each invention as it is made; and thus each day tools are formed or forming, which hereafter, when under moral control, will elevate the material condition of the entire human race. The labour which a hundred years ago made a single shirt now makes a dozen or a score. Ultimately it is possible that the harder and grosser forms of work will be done entirely by machinery, and leisure be left to the human drudge which may lift him bodily into another scale of existence. For the present no such effect is visible. The mouths to be fed and the backs to be covered multiply even faster than the means of feeding and clothing them; and conspicuous as have been the fruits of machinery in the increasing luxuries of the minority, the level of comfort in the families of the labouring millions has in this country been rather declining than rising. The important results have been so far rather political and social. Watt, Stephenson, and Wheatstone, already and while their discoveries are in their infancy, have altered the relation of every country in the world with its neighbours. The ocean barriers between continents which Nature seemed to have raised for eternal separation have been converted into easily travelled highways; mountain chains are tunnelled; distance, once the most troublesome of realities, has ceased to exist. The inventions of these three men determined the fate of the revolt of the Slave States. But for them and their work the Northern armies would have crossed the Potomac in mere handfuls, exhausted with enormous marches. The iron roads lent their help. The collected strength of all New England and the West was able to fling itself into the work; Negro slavery is at an end; and the Union is not to be split like Europe into a number of independent states, but is to remain a single power, to exercise an influence yet unimaginable on the future fortunes of mankind. Aided by the same mechanical faculties, Germany obliterates the dividing lines of centuries. The Americans preserved the unity which they had. The Germans conquer for themselves a unity which they had not. France interferes, and half a millon soldiers are collected and concentrated in a fortnight; armies, driven in like wedges, open rents and gaps from the Rhine to Orleans; and at the end of two months the nation whose military strength was supposed to be the greatest in the world was reeling paralyzed under blows to which these modern contrivances had exposed her. So far we may be satisfied; but who can foresee the ultimate changes of which these are but the initial symptoms? Who will be rash enough to say that they will promote necessarily the happiness of mankind? They are but weapons which may be turned to good or evil, according to the characters of those who best understand how to use them.

The same causes have created as rapidly a tendency no less momentous towards migration and interfusion, which may one day produce a revolution in the ideas of allegiance and nationality. English, French, German, Irish, even Chinese and Hindus, are scattering themselves over the world; some bonâ fide in search of new homes, some merely as temporary residents—but any way establishing themselves wherever a living is to be earned in every corner of the globe, careless of the flag under which they have passed. Far the largest part will never return : they will leave descendants, to whom their connection with the old country will be merely matter of history; but the ease with which we can now go from one place to the other will keep alive an intention of returning, though it be never carried out; and as the numbers of these denizens multiply, intricate problems have already risen as to their allegiance, and will become more and more complicated. The English at Hong Kong and Shanghai have no intention of becoming Chinese, but their presence there has shaken the stability of the Chinese empire, and has cost that country, if the returns are not enormously exaggerated, in the civil wars and rebellions of which they have been the indirect occasion, a hundred million lives.

From the earliest times we trace migrations of nations or the founding of colonies by spirited adventurers; but never was the process going on at such a rate as now, and never with so little order or organized communion of purpose. No ingenuity could have devised a plan for the dispersion of the superfluous part of the European populations so effective as the natural working of personal impulse, backed by these new facilities. The question still returns, however, To what purpose? Are the effects of emigration to be only as the effects of machinery? Are a few hundred millions to be added to the population of the globe merely that they may make money and spend it? In all the great movements at present visible there is as yet no trace of the working of intellectual or moral ideas—no sign of a conviction that man has more to live for than to labour and eat the fruit of his labour.

So far, perhaps, the finest result of scientific activity lies in the personal character which devotion of a life to science seems to produce. While almost every other occupation is pursued for the money which can be made out of it, and success is measured by the money result which has been realized—while even artists and men of letters, with here and there a brilliant exception, let the bankers' book become more and more the criterion of their being on the right road, the men of science alone seem to value knowledge for its own sake, and to be valued in return for the addition which they are able to make to it. A dozen distinguished men might be named who have shown intellect enough to qualify them for the woolsack, or an archbishop's mitre: external rewards of this kind might be thought the natural recompense for work which produces results so splendid; but they are quietly and unconsciously indifferent—they are happy in their own occupations, and ask no more; and that here, and here only, there is real and undeniable progress is a significant proof that the laws remain unchanged under which true excellence of any kind is attainable.

To conclude.

The accumulation of wealth, with its daily services at the Stock Exchange and the Bourse, with international exhibitions for its religious festivals, and political economy for its gospel, is progress, if it be progress at all, towards the wrong place. Baal, the god of the merchants of Tyre, counted four hundred and fifty prophets when there was but one Elijah. Baal was a visible reality. Baal rose in his sun-chariot in the morning, scattered the evil spirits of the night, lightened the heart, quickened the seed in the soil, clothed the hill-side with waving corn, made the gardens bright with flowers, and loaded the vineyard with its purple clusters. When Baal turned away his face the earth languished, and dressed herself in her winter mourning robe. Baal was the friend who held at bay the enemies of mankind—cold, nakedness, and hunger; who was kind alike to the evil and the good, to those who worshipped him and those who forgot their benefactor. Compared to him, what was the being that 'hid himself,' the name without a form—that was called on, but did not answer—who appeared in visions of the night, terrifying the uneasy sleeper with visions of horror? Baal was god. The other was but the creation of a frightened imagination—a phantom that had no existence outside the brain of fools and dreamers. Yet in the end Baal could not save Samaria from the Assyrians, any more than progress and 'unexampled prosperity' have rescued Paris from Von Moltke. Paris will rise from her fallen state, if rise she does, by a return to the uninviting virtues of harder and simpler times. The modern creed bids every man look first to his cash-box. Fact says that the cash-box must be the second concern— that a man's life consists not in the abundance of things that he possesses. The modern creed says, by the mouth of a President of the Board of Trade, that adulteration is the fruit of competition, and, at worst, venial delinquency. Fact says that this vile belief has gone like poison into the marrow of the nations. The modern creed looks complacently on luxury as a stimulus to trade. Fact says that luxury has disorganized society, severed the bonds of good-will which unite man to man, and class to class, and generated distrust and hatred. The modern creed looks on impurity with an appro-bation none the less real that it dares not openly avow it, dreading the darkest sins less than over-population. Fact— which if it cannot otherwise secure a hearing, expresses itself at last in bayonets and bursting shells—declares that if our great mushroom towns cannot clear themselves of pollution, the world will not long endure their presence.

A serious person, when he is informed that any particular country is making strides in civilization, will ask two questions. First personally, Are the individual citizens growing more pure in their private habits? Are they true and just in their dealings? Is their intelligence, if they are becoming intelligent, directed towards learning and doing what is right, or are they looking only for more extended pleasures, and for the means of obtaining them? Are they making progress in what old-fashioned people used to call the fear of God, or are their personal selves and the indulgence of their own inclinations the end and aim of their existence? That is one question, and the other is its counterpart. Each nation has a certain portion of the earth's surface allotted to it, from which the means of its support are being wrung : are the proceeds of labour distributed justly, according to the work which each individual has done; or does one plough and another reap in virtue of superior strength, superior cleverness or cunning?

These are the criteria of progress. All else is merely misleading. In a state of nature there is no law but physical force. As society becomes organized, strength is coerced by greater strength; arbitrary violence is restrained by the policeman; and the relations between man and man, in some degree, are humanized. That is true improvement. But large thews and sinews are only the rudest of the gifts which enable one man to take advantage of his neighbour. Sharpness of wit gives no higher title to superiority than bigness of muscle and bone. The power to overreach requires restraint as much as the power to rob and kill; and the progress of civilization depends on the extent of the domain which is reclaimed under the moral law. Nations have been historically great in proportion to their success in this direction. Religion, while it is sound, creates a basis of conviction on which legislation can act; and where the legislator drops the problem, the spiritual teacher takes it up. So long as a religion is believed, and so long as it retains a practical direction, the moral idea of right can be made the principle of government. When religion degenerates into superstition or doctrinalism, the statesman loses his ground, and laws intended, as it is scornfully said, to make men virtuous by Act of Parliament, either sink into desuetude or are formally abandoned. How far modern Europe has travelled in this direction would be too large an inquiry. Thus much, however, is patent, and, so far as our own country is concerned, is proudly avowed : Provinces of action once formally occupied by law have been abandoned to anarchy. Statutes which regulated wages, statutes which assessed prices, statutes which interfered with personal liberty, in the supposed interests of the commonwealth, have been repealed as mischievous. It is now held that beyond the prevention of violence and the grossest forms of fraud, government can meddle only for mischief—that crime only needs repressing—and that a community prospers best where every one is left to scramble for himself, and find the place for which his gifts best qualify him. Justice, which was held formerly to be co-extensive with human conduct, is limited to the smallest corner of it. The labourer or artisan has a right only to such wages as he can extort out of the employer. The purchaser who is cheated in a shop must blame his own simplicity, and endeavour to be wiser for the future.

Habits of obedience, moral convictions inherited from earlier times, have enabled this singular theory to work for a time; men have submitted to be defrauded rather than quarrel violently with the institutions of their country. There are symptoms, however, which indicate that the period of for-bearance is waning. Swindling has grown to a point among us where the political economist preaches patience unsuccessfully, and Trades-Unionism indicates that the higgling of the market is not the last word on the wages question. Government will have to take up again its abandoned functions, and will understand that the cause and meaning of its existence is the discovery and enforcement of the elementary rules of right and wrong. Here lies the road of true progress, and nowhere else. It is no primrose path—with exhibition flourishes, elasticity of revenue, and shining lists of exports and imports. The upward climb has been ever a steep and thorny one, involving, first of all, the forgetfulness of self, the worship of which, in the creed of the economist, is the mainspring of advance. That the change will come, if not to us in England, yet in our posterity somewhere upon the planet, experience forbids us to doubt. The probable manner of it is hopelessly obscure. Men never willingly acknowledge that they have been absurdly mistaken.

An indication of what may possibly happen can be found, perhaps, in a singular phenomenon of the spiritual develop-ment of mankind which occurred in a far distant age. The fact itself is, at all events, so curious that a passing thought may be usefully bestowed upon it.

The Egyptians were the first people upon the earth who emerged into what is now called civilization. How they lived, how they were governed during the tens of hundreds of generations which intervened between their earliest and latest monuments, there is little evidence to say. At the date when they become distinctly visible they present the usual features of effete Oriental societies; the labour executed by slave gangs, and a rich luxurious minority spending their time in feasting and revelry. Wealth accumulated, Art flourished. Enormous engineering works illustrated the talent or ministered to the vanity of the priestly and military classes. The favoured of fortune basked in the perpetual sunshine. The millions sweated in the heat under the lash of the task-master, and were paid with just so much of the leeks and onions and fleshpots as would continue them in a condition to work. Of these despised wretches some hundreds of thousands were enabled by Providence to shake off the yoke, to escape over the Red Sea into the Arabian desert, and there receive from heaven a code of laws under which they were to be governed in the land where they were to be planted.

What were those laws?

The Egyptians, in the midst of their corruptions, had inherited the doctrine from their fathers which is considered the foundation of all religion. They believed in a life beyond the grave—in the judgment bar of Osiris, at which they were to stand on leaving their bodies, and in a future of happiness or misery as they had lived well or ill upon earth. It was not a speculation of philosophers—it was the popular creed; and it was held with exactly the same kind of belief with which it has been held by the Western nations since their conversion to Christanity.

But what was the practical effect of their belief? There is no doctrine, however true, which works mechanically on the soul like a charm. The expectation of a future state may be a motive for the noblest exertion, or it may be an excuse for acquiescence in evil, and serve to conceal and perpetuate the most enormous iniquities. The magnate of Thebes or Memphis, with his huge estates, his town and country palaces, his retinue of eunuchs, and his slaves whom he counted by thousands, was able to say to himself, if he thought at all, ' True enough, there are inequalities of fortune. These serfs of mine have a miserable time of it, but it is only a *time* after all; they have immortal souls, poor devils! and their wretched existence here is but a drop of water in the ocean of their being. They have as good a chance of Paradise as I have—perhaps better. Osiris will set all right hereafter; and for the present rich and poor are an ordinance of Providence, and there is no occasion to disturb established institutions. For myself, I have drawn a prize in the lottery, and I hope I am grateful. I subscribe handsomely to the temple services. I am myself punctual in my religious duties. The priests, who are wiser than I am, pray for me, and they tell me I may set my mind at rest.'

Under this theory of things the Israelites had been ground to powder. They broke away. They too were to become a nation. A revelation of the true God was bestowed on them, from which, as from a fountain, a deeper knowledge of the Divine nature was to flow out over the earth; and the central thought of it was the realization of the Divine government— not in a vague hereafter, but in the living present. The unpractical prospective justice which had become an excuse for tyranny was superseded by an immediate justice in time. They were to reap the harvest of their deeds, not in heaven, but on earth. There was no life in the grave whither they were going. The future state was withdrawn from their sight till the mischief which it had wrought was forgotten. It was not denied, but it was veiled in a cloud. It was left to private opinion to hope or to fear; but it was no longer held out either as an excitement to piety or a terror to evil-doers. The God of Israel was a living God, and His power was displayed visibly and immediately in rewarding the good and punishing the wicked while they remained in the flesh.

It would be unbecoming to press the parallel, but phenomena are showing themselves which indicate that an analogous suspension of belief provoked by the same causes may possibly be awaiting ourselves. The relations between man and man are now supposed to be governed by natural laws which enact themselves independent of considerations of justice. Political economy is erected into a science, and the shock of our moral nature is relieved by reflections that it refers only to earth, and that justice may take effect hereafter. Science, however, is an inexorable master. The evidence for a hereafter depends on considerations which science declines to entertain. To piety and conscientiousness it appears inherently probable; but to the calm, unprejudiced student of realities, piety and conscientiousness are insufficient witnesses to matters of fact. The religious passions have made too many mistakes to be accepted as of conclusive authority. Scientific habits of thought, which are more and more controlling us, demand external proofs which are difficult to find. It may be that we require once more to have the living certainties of the Divine government brought home to us more palpably; that a doctrine which has been the consolation of the heavy-laden for eighteen hundred years may have generated once more a practical infidelity; and that by natural and intelligent agencies, in the furtherance of the everlasting purposes of our Father in heaven, the belief in a life beyond the grave may again be about to be withdrawn.

INDEX